Combustion of Pulverised Coal in a Mixture of Oxygen and Recycled Flue Gas

Combustion of Pulverised Coal in a Mixture of Oxygen and Recycled Flue Gas

Dobrin D. Toporov

AMSTERDAM • BOSTON • HEIDELBERG • LONDON • NEW YORK • OXFORD
PARIS • SAN DIEGO • SAN FRANCISCO • SINGAPORE • SYDNEY • TOKYO

ELSEVIER

Elsevier
225 Wyman Street, Waltham, MA 02451, USA
32 Jamestown Road, London NW1 7BY

Notices
Knowledge and best practice in this field are constantly changing. As new research and experience broaden our understanding, changes in research methods, professional practices, or medical treatment may become necessary.

Practitioners and researchers must always rely on their own experience and knowledge in evaluating and using any information, methods, compounds, or experiments described herein. In using such information or methods they should be mindful of their own safety and the safety of others, including parties for whom they have a professional responsibility.

To the fullest extent of the law, neither the Publisher nor the authors, contributors, or editors, assume any liability for any injury and/or damage to persons or property as a matter of products liability, negligence or otherwise, or from any use or operation of any methods, products, instructions, or ideas contained in the material herein.

British Library Cataloguing in Publication Data
A catalogue record for this book is available from the British Library

Library of Congress Cataloging-in-Publication Data
A catalog record for this book is available from the Library of Congress

ISBN: 978-0-08-099998-2

For information on all **Elsevier** publications
visit our web site at store.elsevier.com

This book has been manufactured using Print On Demand technology. Each copy is produced to order and is limited to black ink. The online version of this book will show color figures where appropriate.

Working together
to grow libraries in
developing countries

www.elsevier.com • www.bookaid.org

Contents

List of Figures

List of Tables

Acknowledgments

The work presented here is based on the work I conducted as a post-doctoral fellow at the Institute of Heat and Mass Transfer at RWTH Aachen University between April 2004 and December 2010.

I am especially indebted to Prof. Dr.-Ing. Reinhold Kneer (Chair of the Institute of Heat and Mass Transfer, RWTH Aachen University) for his professional and personal support and in particular for having constant confidence in my work.

I owe my gratitude to Dr. Joao Luis Toste de Azevedo for his willingness to act as assessor. Dr. Azevedo gave me the opportunity (after finishing my Ph.D. at the Technical University at Sofia) to work with several interesting research projects at the Instituto Superior Tecnico, Lisbon, and he contributed to the success of my work via his expertise in the field of coal combustion modelling. I will always remember the fruitful exchange of ideas and warm hospitality I experienced there. I am very grateful to Prof. Dr.-Ing. Roman Weber (Technical University Clausthal, Germany) and Dr. Kurose (Kyoto University, Japan) for agreeing to be reviewers and for their continued interest in this work.

I would like to thank to Prof. Ulrich Renz for giving me the opportunity to work at WSA, RWTH Aachen. My thanks also to Dr. Bernd Hillemacher for his professional and personal support during my stay in Aachen.

My special thanks to Dr. Malte Förster, who created the framework for my study and who gave me every support to ensure success in my work. In Dr. Förster I found a stimulating discussion partner on basic concepts of oxycoal combustion chemistry. He was always willing to listen to scientifically related difficulties and was always helpful in finding solutions when problems arose.

I would like to thank all my former colleagues and all those who were involved directly and indirectly in this work at the Institute of Heat and Mass Transfer. The experimental results reported in this study would not be possible without their support and their strong engagement in the oxycoal test facility.

The financial assistance from the German Federal Ministry of Economics and Technology (BMWi, fkz 0326890a), Ministry of Innovation, Science, Research and Technology of the State of NRW (miwft, fkz 22320601703), RWE Power AG, E.on AG, Siemens AG, Linde AG, Hitachi Power Europe GmbH, MAN Turbo AG, and WS-Wearmeprozesstechnik GmbH, under the project OXYCOAL-AC, is also acknowledged.

Finally, I want to express the biggest thanks to my wife Mariana for her patience as well as her unlimited understanding and support, without which such work would not have been possible.

Nomenclature

Symbol

Symbol	Description	Units
A	Area	$[m^2]$
A	Pre-exponential factor	[varies]
A_k	Mole fraction	[–]
A, B	Empirical coefficients in EDM	[–]
c	Molar concentration	$[mol/m^3]$
C	Function, coefficient, constant	[–]
C_{D1}, C_{D2}	Empirical coefficients in EDC	[–]
c_p	Specific heat capacity	[J/kg-K]
d	Diameter	[m]
D	Coefficient	[–]
E	Activation energy	[kJ/mol]
f	Force	[N]
ΔH	Heat of formation	[J/kg]
h	Heat transfer coefficient	$[W/m^2\text{-}K]$
I	Radiation intensity	$[W/m^2\text{-}sr]$
k	Turbulent kinetic energy	$[m^2/s^2]$
K	Constant	[–]
L	Characteristic length	[m]
m, M	Mass	[kg]
M	Molecular weight	[kg/mol]
m	Mass flow rate	[kg/s]
p, P	Pressure	[Pa]
p	Ratio	[–]
Q	Thermal input	[W]
r	Stoichiometric ratio	[–]
R	Reaction rate	$[kg/(m^3\ s)]$
R	Universal gas constant	8,314 kJ/(kmol K)
S	Particle internal surface area	$[m^2]$
S	Source terms	[varies]
t	Time	[s]
T	Temperature	[K], [°C]
u, v	Velocity components	[m/s]
u', v'	Fluctuation of velocity	[m/s]
V	Mass	[kg]
X	Distance	[m]
Y	Mass fraction	[–]

Greek Letters

α	Absorptivity	[–]
γ	Mass fraction	[–]
ε	Emissivity	[–]
ε	Effectiveness factor	[–]
ε	Dissipation rate of turbulent energy	$[m^2/s^3]$
ϕ	Source term	[varies]
ϕ	Ratio	[–]
Γ	Diffusion coefficient	$[kg/(m\,s)]$
δ, Δ	Difference operator	[–]
δ	Particle pore diameter	$[10^{-9}\,m]$
η	Dynamic viscosity	$[kg/(m\,s)]$
λ	Excess air ratio	[–]
ν	Kinematic viscosity	$[m^2/s]$
ν	Stoichiometric coefficient	[–]
ν	Frequency	$[1/s]$
ρ	Density	$[kg/m^3]$
ρ	Reaction-rate ratio in CPD model	[–]
τ	Time constant	$[s]$
τ	Tortuosity	[–]
τ	Shear stresses	$[kg/(s^2\,m)]$
σ	Model constants	[–]
σ	Stefan-Boltzmann constant	$5,669\ W/(m^2\,k)$
ω	Reaction rate	$[1/s]$

Abbreviations

AFT	Adiabatic Flame Temperature
ASTM	American Society for Testing and Materials
BFB	Bubbling Fluidised Bed
BO	Burnout
d.a.f.	Dry Ash-Free
DIN	Deutsches Institut fuer Normung
CBK	Char Burnout Kinetics Model
CCS	Carbon Capture and Storage
CCR	Carbon Capture and Reuse
CDL	Coal Devolatilisation Laboratory
CFB	Circulating Fluidised Bed
CFD	Computational Fluid Dynamics
CPD	Chemical Percolation Devolatilisation Model
CR	Conversion Ratio
EDC	Eddy Dissipation Concept
EDM	Eddy Dissipation Model
ER	External Recirculation
ESP	Electrostatic Precipitators

EWBM	Exponential Wideband Model
FB	Fluidised Bed
FGD	Flue Gas Desulphurisation
FG-DVC	Functional Group, Depolymerisation, Vaporisation, and Cross-Linking
GHG	Greenhouse Gas
GCV	Gross Calorific Value
IGCC	Integrated Gasification Combine Cycle
IPFR	Isothermal Plug Flow Reactor
IRZ	Internal Recirculation Zone
KIN	Kinetics
LES	Large Eddy Simulation
L-H	Langmuir-Hinshelwood
max	Maximum
min	Minimum
Mtoe	Million (Mega-) Tonnes of Oil Equivalent
NMR	Nuclear Magnetic Resonance
OECD	Organisation for Economic Cooperation and Development
OFA	Overfire Air
PFBC	Pressurised Fluidised Bed Combustion
PFR	Plug Flow Reactor
PM	Primary Stream Momentum Ratio
ppm	Parts Per Million
PS	Primary Stream
RANS	Reynolds Averaged Navier Stokes
RFG	Recirculated Flue Gas
R/P	Reserves-to-Production
SCR	Selective Catalytic Reduction
SK	Steinkohle
SM	Secondary Stream Momentum Ratio
SS	Secondary Stream
TA	Thermal Annealing
TGA	Thermogravimetric Analysis
TS	Tertiary Stream
VM	Volatile Matter
Vol	Volume
URANS	Unsteady Reynolds Averaged Navier Stokes
WSGGM	Weighted Sum of Grey Gases Model
wt	Weight
WTA	Wirbelschicht-Trocknung von Braunkohle mit interner Abwärmenutzung

Dimensionless Numbers

Nu	Nusselt number
Re	Reynolds number
Sh	Sherwood number

Subscripts

ad	Adsorption
ash	Related to fuel ash
app, a	Apparent
b	Bulk
b	Burner
c	Related to fuel carbon
ch	Chemical
char	Related to fuel char
D	Diffusion
f	Forward
f	Fluid
fuel	Fuel
h	Heat
i	Summation index
k	Gas specie index or summation index
l	Reaction index or summation index
m	Mean
o	Initial
p	Particle
prod	Product
s	Scattering or surface
t	Turbulence
w	Wall
∞	Free stream values

Superscripts

$'$	Reactant
$''$	Product
$''$	Fluctuating part
$*$	Fine structure in EDC model
o	Surrounding in EDC model
k	Particle class index

1 Introduction

During the last century, the world witnessed an accelerating technological development in almost all aspects of human life, resulting in rapidly improving living standards in the vast majority of countries. This development would have been impossible without cheap and available energy, and the growing demand for energy services has led to both the discoveries of new energy sources and the development of new energy conversion technologies.

Overall, it seems probable that the dominance of the main commercial fossil fuels will continue into the foreseeable future, while several key "new" renewable energies (e.g., wind, solar, biomass, etc.) seem to be on the path of rapid growth and declining cost. Renewable energy will undoubtedly become an integral and important component of the future energy mix, but major expansion on a global scale will take time. In the foreseeable future, renewables can only complement conventional energies. From the other side, the high risks related to the utilisation of nuclear technology for energy production lead to many societies' growing lack of acceptance of this technology.

Coal has played and continues to play a significant role in meeting the global demand for energy services. Coal, being the most abundant, available, and affordable fuel, has the potential to become the most reliable and easily accessible energy source, thus able to make a crucial contribution to world energy security.

The major challenges facing coal, however, are concerned with its environmental impacts in terms of both its production and its use. Nowadays, pulverised coal (PC)-fired power plants demonstrate high reliability and availability and are much cleaner than ever before. Emissions of NO_x, SO_2, and particulates are reduced by over 90% on many older plants relative to uncontrolled levels. This is accomplished by advanced combustion and backend cleanup systems.

More recently, greenhouse gas (GHG) emissions, including carbon dioxide (CO_2), have become a concern because of their possible relations to climate change. A number of options exists to reduce CO_2 emissions. Recently, CO_2 capture and storage technologies applied to the coal-based electricity and heat generation sector, being among the major sources of CO_2, have gained huge interest as a promising option that has the potential to reduce these emissions drastically. This concept is usually divided into three different approaches: post-combustion capture, pre-combustion capture, and oxyfuel combustion capture. The current work is focused on the oxyfuel pulverised coal combustion because of easy CO_2 recovery and low NO_x emissions.

The combustion of pulverised fuel in a mixture of recycled flue gas (RFG) and oxygen, however, presents new challenges to combustion specialists. Several experimental investigations with oxy-firing pulverised coal burners report that flame temperature and stability are strongly affected [1–5]. Utilising a burner design that was optimised for coal combustion in air will lead to flame instability and poor burnout for oxycoal combustion. In addition, in contrast to air-blown systems, oxy-firing provides the unique possibility to vary a whole set of parameters, such as temperature level, the oxygen concentration used for firing, and the composition of recycled flue gas. Hence, in order to obtain optimum process conditions, basic research on pulverised fuel oxy-firing, combined with bench- and pilot-scale experiments, is required.

Starting from the knowledge that carbon dioxide exhibits pronounced differences in thermodynamic and optical properties compared to air, the primary objective of this work is to investigate the effects of the particular properties of CO_2 on the chemical reactions and on the heat transfer that takes place during oxycoal combustion. The next stage of this study aims at the formulation of basic principles for control of the process of PF combustion in CO_2/O_2 atmosphere, thus enabling its subsequent implementation.

First, in Chapter 2, the state-of-the-art coal utilisation technologies are presented. A detailed review of the achievements in the field of clean coal technologies is made with respect to the most recent developments in an efficient and environmentally friendly use of coal. Finally, the oxyfuel technology is introduced, and the main challenges to be faced before its implementation on a large scale are set down.

This work is divided into two main parts:

Part I (Chapters 3 and 4) considers the theoretical aspects of the process of combustion of pulverised fuel (PF) in a mixture of RFG and O_2. Chapter 3 highlights the basic mechanisms of coal particle combustion, with particular attention given to combustion in an atmosphere with high content of CO_2 and water vapour. The main differences in burning PF in air and in RFG/O_2 atmospheres are identified, and the resulting impacts on the characteristics of the flame and on the design of PF burners are addressed in Part II of this study. Further, the modelling of PF combustion as a general concept and the specifics in modelling oxyflames, in particular, are addressed in Chapter 4. A CFD-based numerical model, including submodels for homogeneous and heterogeneous combustion that are adapted for oxyfuel flame modelling, is introduced. The possible drawbacks and advantages of the existing models for turbulence, turbulence-chemical reactions interactions, coal particle devolatilisation, and char burnout as well as for gas emissivity are discussed and recommendations for their use are made. A detailed validation and assessment of the capabilities of state-of-the-art coal particle combustion model approaches covering different atmospheres and a variety of conditions is performed.

Part II (Chapters 5 and 6) summarises the main results obtained during a scientific investigation on oxyfuel pulverised coal combustion, including theoretical as well as bench- and pilot-scale experiments that have been carried out at the Institute of Heat and Mass Transfer at the RWTH Aachen University, Germany.

Chapter 5 compiles the experimental and numerical investigations of combustion of gaseous fuel in N_2/O_2 and CO_2/O_2 atmospheres, performed in a 25 kW gas reactor. Here the effects of CO_2 on the oxidation rates of methane and carbon monoxide are investigated and analysed.

Further, Chapter 6 presents the development, design, testing, and scale-up of a swirl burner for oxyfuel combustion of pulverised coal. Detailed data of the oxycoal flame characteristics, obtained experimentally by varying burner operating conditions at the

100 kW oxycoal test facility, are provided with the purpose of model validations. The effects of the composition of RFG, of the recycling ratio (O_2 content, respectively), and of the burner settings on the flame stability and carbon conversion are analysed. Thus the underlying principles for design and operation of large-scale burners able to operate in both air and oxyfuel PF combustion modes are formulated.

Finally, Chapter 7 deals with the NO_x emissions in oxyfuel pulverised coal combustion and presents an assessment of the influence of different operational parameters on NO_x conversion in a real oxycoal swirl flame.

2 Coal Combustion Technologies

2.1 Coal demand and coal reserves worldwide

During the last three decades, primary energy consumption has increased worldwide by about 70% (Figure 2.1), reaching 11 gigatonnes oil equivalent (Gtoe) at the end of 2009. There was a fast increase in oil and natural gas consumption, with shares of total consumption at 35% and 25%, respectively.

After some stagnation up to 2002, coal consumption in the world in 2009 increased rapidly for the seventh consecutive year. Global consumption rose 4.5%, a rate above the 10-year average of 3.2%. Consumption growth was widespread, but the sharp increase in coal consumption after 2002 is related mainly to the increased coal consumption in China. After growing at an average rate of 3% per year from 1990 to 2001, China's coal consumption on average increased 17% per year from 2002 to 2005. As a result, coal usage in China has nearly doubled since 2000, and, given the country's rapidly expanding economy and large domestic coal deposits, its demand for coal is projected to continue growing strongly.

In the International Energy Agency's (IEA's) *New Policies Scenario* [6], world primary energy demand will increase 36% between 2008 and 2035, from around 12.3 Gtoe to over 16.7 Gtoe, or a 1.2% increase per year on average. Fossil fuels, oil, coal, and natural gas are predicted to remain the dominant energy sources in 2035.

Global coal demand growth under the *New Policies Scenario* will be around 20% between 2008 and 2035, with 100% of this increase occurring in non-Organisation for Economic Cooperation and Development (OECD) countries. Global coal demand is expected to peak around 2025 and begin to decline, slowly returning to 2003 levels by 2035 due to the expected constranency by climate policy measures.

Globally, coal will remain the leading source of electricity generation in 2035, although its share of electricity generation will decline from 41% to 32%. Nonetheless, total coal demand is projected to reach 5.6 billion tonnes (or 3.9 Mtoe) in 2035, up from 4.7 billion tonnes (or 3.3 Mtoe) in 2008 [6].

The production of all fossil fuels at the end of 2009 totals 457 EJ, or 10.9 Mtoe (according to the German Federal Institute for Geosciences and Natural Resources [BGR], [7]), and 440 EJ, or 10.5 Mtoe (according to BP Statistical Review of World Energy, [8]). Reserves amount to 39, 794 EJ, or 951 Mtoe, and resources are in the order of 613, 180 EJ, or 14, 655 Mtoe. The ratios of annual production to reserves and to resources are about 1 to 87 and 1 to 1342, respectively, showing that the world's energy demand can be covered for quite some time [7,8].

Combustion of Pulverised Coal in a Mixture of Oxygen and Recycled Flue Gas. http://dx.doi.org/10.1016/B978-0-08-099998-2.00002-3

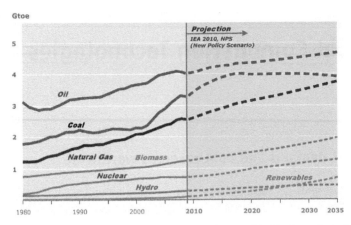

Figure 2.1 Development of primary energy consumption worldwide (cumulative) by energy type, 1980–2010, and IEA projections to 2035 [6].

Coal possesses the largest potential of all nonrenewable fuels and provides 53% of the reserves and 77% of the resources worldwide. The remaining potential of hard coal and lignite is sufficient to cover the expected demand for many decades to come [7].

Coal is an inhomogeneous organic fuel, formed largely from partially decomposed and metamorphosed plant materials. Formation has occurred over long time periods, often under high pressures of overburden and at elevated temperature. Differences in plant materials and in their extent of decay influence the components present in coal. Therefore, coal types vary greatly in their composition. Over the years, many efforts have been made to sort and classify the almost limitless number of coal types in broad classifications and to relate similarities among coal types to their potential behaviour in the coal conversion process. These efforts resulted in many classification systems currently in use in the member countries of the IEA, such as the American Society for Testing and Materials, or ASTM (used in North America), the National Coal Board, or NCB (UK), the Australian system (including the new 1987 system), and the German (Ruhr) and UN-ECE (international) (for both hard and soft coal) classifications. Table 2.1 presents some of these coal classifications.

Recently, torrefied biomass, processed at temperatures in the range of 270–320 °C, became an attractive fuel for co-combustion in PF firing boilers. Although there is still no classification system for this type of fuel, typical lower heating values (LHVs) are in the range of 21–26 MJ/kg, with high volatiles content ($\approx$70%), low ash content, and very reactive char [9].

Although questions regarding the size and location of reserves of oil and gas raise increasing concern, coal remains abundant and broadly distributed around the world. Economically recoverable reserves of coal are available in more than 70 countries

Table 2.1 Coal classifications referring to ultimate analysis and calorific value on a dry mineral-matter-free basis.

Coal Types			Volatiles	C	H	O	S	GCV
Germany (DIN)	UN-ECE	USA (ASTM)	%	%	%	%	%	kJ/kg
Braunkohle	Lignite	Lignite	≥45	≤75	6–5.8	34–17	0.5–3	≤22,100
Matt-und Glanz-Braunkohle	Sub bituminous	Sub bituminous	≥45	≤75	6–5.8	34–17	0.5–3	≤28,470
Flamm-SK		High volatile bituminous	40–45	75–82	6–5.8	≤9.8	~1	≤32,870
Gasflamm-SK		bituminous	35–40	82–85	5.8–5.6	9.8–7.3	~1	≤33,910
Gas-SK	Bituminous	Medium volatile	28–35	85–87	5.6–5	7.3–4.5	~1	≤34,960
Fett-SK		bituminous	19–28	87–89	5–4.5	4.5–3.2	~1	≤35,380
Ess-SK		Low volatile bituminous	14–19	89–90.5	4.5–4	3.2–2.8	~1	35,380
Mager-SK	Anthracite	Semi-anthracite	10–14	90.5–91.5	4–3.7	2.8–2.5	~1	≤35,380
Anthrazit		Anthracite	7–12	≤91.5	≤3.7	≤2.5	~1	≤35,300

Table 2.2 Proven coal reserves at end of 2009 in million tonnes, according to BP statistics [8](first 12 countries).

Country	Antracite and Bituminous	Subbituminous and Lignite	Total	Share of Total (%)	R/P Ratio in Years
United States	108,950	129,358	238,308	28.9	245
Russian Federation	49,088	107,922	157,010	19.0	≥500
P.R. China	62,200	52,300	114,500	13.9	38
Australia	36,800	39,400	76,200	9.2	186
India	54,000	4,600	58,600	7.1	105
Ukraine	15,351	18,522	33,873	4.1	460
Kazakhstan	28,170	3,130	31,300	3.8	308
South Africa	30,408	. . .	30,408	3.7	122
Poland	6,012	1,490	7,502	0.9	56
Brazil	. . .	7,059	7,059	0.9	≥500
Colombia	6,434	380	6,814	0.8	95
Germany	152	6,556	6,708	0.8	37
. . .	. . .	. . .	. . .	. . .	. . .
World	411,321	414,680	826,001	100	119

worldwide and in each major world region. Proven[1] reserves at the end of 2009 amounted to around 826 billion tonnes (the geological resource is far larger), equivalent to 119 years at current production rates[2] [8].

Coal is found in many countries, but more than 90% of reserves are located in just eight countries: the United States (28.9%), the Russian Federation (19.0%), China (13.9%), Australia (9.2%), India (7.1%), South Africa (3.7), Ukraine (4.1%), and Kazakhstan (3.8%), as shown in Table 2.2. Many other countries hold large reserves as well.

The ratio of reserves to annual coal production (R/P) for Europe (OECD) and Germany is 50 and 37, respectively. When the increase of coal production is considered, however, the ratio becomes lower.[3]

2.2 Coal utilisation processes

Most of the coal now being consumed worldwide is by direct combustion of finely pulverised coal in large-scale utility furnaces for electric power generation. However,

---

[1]Proven reserves of coal: Generally taken as those quantities that geological and engineering information indicates with reasonable certainty to be recoverable in the future from known deposits under existing economic and operating conditions.

[2]Reserves-to-production (R/P) ratio: If the reserves remaining at the end of the year are divided by the production of that year, the R/P ratio is the length of time that those remaining reserves would last if production were to continue at that rate.

[3]The data concerning Germany's and Poland's coal reserves varies, depending on the data source. The difference comes from the different methods of estimation.

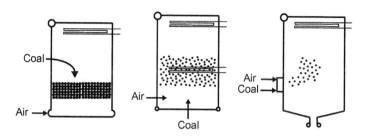

Figure 2.2 Schemes of fixed bed (left), fluidised bed (middle), and entrained flow (right) boilers.

there are many other processes for the conversion of coal into products or for direct combustion. Some of the more common processes will be briefly reviewed here.

Coal combustion or gasification methods include fixed bed, fluidised bed, and entrained flow methods, as shown schematically in Figure 2.2. The grain size of coal becomes progressively smaller in this order, whereas the gas flow velocity in the equipment increases.

In a fixed bed process, the input block coal is fed-pushed, dropped, or thrown on a slowly moving bed of coal particles on a grate. The raw coal is heated, dried, devolatilised, and burned, leaving ash. The primary air flows upwards through the grate and through the bed of coal particles. The combustion gases formed in the bed leave the reactor (in the case of gasification) or are burned (in the case of combustion) in the space above the bed by introducing secondary or over-fire air. Thus, coal reacts statically with air, almost independent from the flow of gases and with increased dependence of the reaction rates on diffusion. The method has the advantage of using block coal directly and burning coals of a wide range of coal rank (from anthracite to lignite), but the boiler efficiency remains low, suppressed by the high excess air ($\approx$40%) that is required for acceptable coal burnout. Furthermore, the method makes the upscaling of a boiler difficult and has limitations in absolute scale.

In the fluidised bed process, crushed coal of 5–10 mm diameter is injected into a hot layer of inert solids (diameter: 0.5–3.0 mm) fluidised with airflow and is combusted or gasified within this layer. The coal particles make up only around 1–2% of the bed mass; the rest are coal ash and limestone or dolomite, which are added to capture sulphur in the course of combustion. In some cases inert materials are also added as bed material. At low air velocities, the air flows through the bed without disturbing the particles, and the bed remains fixed. At velocities greater than the minimum fluidisation velocity, the bed is fluidised and the air flows through the bed in bubbles, thus creating a bubbling fluidised bed (BFB). At velocities approaching or greater than the free-fall velocity of the particles, the particles become entrained in the air and are carried out of the furnace. The entrained particles are separated from the combustion gases in a cyclone and circulated back to the bed, thus creating a circulating fluidised bed (CFB). This recirculation provides large particle residence times in the CFB combustor and allows combustion to take place at a lower temperature. The longer residence times increase the ability to efficiently burn high-moisture, high-ash, low-reactivity, and other hard-to-burn fuel such as anthracite, lignite, and waste coals. Thus with a given design, a wide range of fuels can be burnt.

This process can be realised at atmospheric and pressurised conditions. The coal particles, introduced in the fluidised bed, heat up, dry, devolatilise, ignite, and burn, leaving ash. The residence time of the coal particles in the bed is typically around 1 min and usually sufficient for 80–90% burnout. The bed is cooled by steam-generating tubes immersed in the bed to a temperature in the range of 1050–1170 K. This prevents the softening of the coal ash and the decomposition of $CaSO_4$, the product of sulphur capture. Adding MgO or Al_2O_3 can also affect the ash melting temperature and thus can prevent particle agglomeration inside the fluidised bed. The FB and CFB are mature technologies primarily applied to the use of low-grade fuels in smaller unit sizes. A limitation for scale-up is the FB coal feed system: A feed point is needed for about every 3 MW_{el} output [10] in order to ensure uniform coal distribution in the bed. The advantage of CFB over FB is that it requires fewer feed points due to the smaller cross-sectional area of the bed, and hence it is easier to scale to higher inputs. The CFB technology provides better heat transfer within the fluidised bed, so the size of the boiler can be reduced. Furthermore, the combustion at low temperatures of 800–900 °C prevents formation of thermal nitric oxides and helps in reducing fuel nitric oxides. This also helps reduce slagging and fouling and thus the dependence on coal ash properties. The method can apply in-furnace desulphurisation by injection of fine-grained limestone, which is calcined in the combustor to form calcium oxide. This calcium oxide reacts with sulphur dioxide gas to form solid calcium sulphate. Depending on the fuel and site requirements, additional NO_x and SO_2 reduction measures can be added to the exhaust gases. With this combination of environmental controls, CFB technology provides an excellent option for low emissions and very fuel-flexible power generation. The method, however, has some issues to be solved, such as increased erosion of boiler tubes and boiler walls by fluidised particles. Caution should be paid to reuse of coal ash mixed with gypsum by in-furnace desulphurisation. CFB technology has been an active player in the power-generation market for the last two decades. Today, over 50, 000 MW_{th} of CFB plants are in operation worldwide [11].

In the entrained flow systems, pulverised fine coal flows with air and is burned or gasified in a reactor. The coal is crushed, dried, and pulverised to a fine powder. Typically, 80–90% of the pulverised coal particles pass through a 200-mesh sieve of 74 μm aperture. The pulverised coal is transported with a small part (15–25%) of the total air needed for combustion, the primary air, to a burner and into a furnace. Its temperature is limited to about 100 °C for reasons of safety against ignition and explosion in the mill and the transport pipeline between the mill and the burners. A greater part of the total air, the secondary air, is usually strongly preheated and introduced through the burner ports into the furnace to enable the combustion. The particles devolatilise, ignite, and burn, leaving ash; the residence time of the pulverised coal particles in the furnace is typically 1–2 sec. and is usually sufficient for nearly complete combustion.

The pulverised coal burners used in the entrained flow systems are of two main types: swirl burners and jet burners.

Swirl burners are double concentric, with a central flow containing the coal and primary (transporting) flow and an annular hot secondary (combusting) flow with an expansion at the nozzle (quarl) that generally accommodates the jet expansion. The coal and primary air as well as the secondary air are introduced into the furnace with a strong, swirling rotation. The burner geometry as well as the swirl level determine the flow and mixing patterns, which then determine coal ignition. Swirl burners are usually mounted on the furnace walls, with burner axis being normal to the walls. Each burner is largely independent and has its own flame envelope. Swirl burners are used mostly for burning bituminous coals.

In the jet burner arrangements, the coal and primary stream as well as the secondary stream are introduced into the furnace as jets from vertical nozzle arrays with no rotation. Burners of this type are often placed at the corners of the furnace so that the coal and the air jets are tangential to an imaginary vertical cylinder in the middle of the combustion chamber. This creates a large vortex in the center of the furnace. Jet burners operate together, creating a single flame envelope. They are usually used for coals with high moisture content, such as lignite. Recently, technologies for pre-drying of high-moisture, low-quality coal using waste heat in the form of low-temperature steam (WTA drying) [12] or hot air [13] in a fluidised bed were developed, thus enabling the utilisation of lignite in wall-fired furnaces at an improved thermal efficiency.

Traditionally, low-volatile coals such as semi-anthracites and anthracites have been utilised in arch-fired furnaces (often referred to as "downshot"-fired furnaces) so as to overcome the inherent difficulties of these low-reactive fuels. The downshot firing system is, however, more expensive than a comparable wall-fired plant. Therefore, significant efforts have been made recently to burn low-volatile coals in wall-fired, horizontal furnaces by adjusting the design and operation conditions of conventional swirl burners [14].

Pulverised coal-fired furnaces are usually classified according to the method of ash removal into molten-ash and dry-ash furnaces.

In molten-ash furnaces, the boiler tubes in the lower part of the furnace are covered by a refractory to reduce heat extraction and to allow the combustion temperature to rise beyond the melting point of the ash. The temperature is normally sufficiently high for the viscosity of the slag to be reduced to about 150 poise,[4] which is necessary for removal in liquid form. Because of the high temperature and the reducing atmosphere in molten-ash furnaces, they are characterised with very high NO_x emissions and high temperature corrosion, respectively.

In dry-ash furnaces, most of the fly ash formed in pulverised coal combustion is removed from the flue gas in the form of dry particulate matter, with a small proportion (10%) of the coal ash falling off the tube walls as semi-molten agglomerated ash, which is collected from the bottom hopper of the combustion chamber. Although there are some difficulties with handling and disposing of dry fly ash, this type of furnace is much more common in use, since it is simpler, more flexible, and more reliable in comparison with the molten-ash furnace [15].

The main advantages of pulverised coal combustion are high reliability, full automation, adaptation to a wide range of coal ranks and operating requirements, excellent capacity for increasing unit size, and cost-effective power generation. The main disadvantages are high energy consumption for pulverising coal, high particulate emissions,

---

[4]1 poise = 0.1 Ns/m^2.

SO_2 and NO_x emissions. Over 2 million MW_{th} of pulverised coal power plants are operating globally [16].

In addition to the coal utilisation technologies we have mentioned, other options, such as coal gasification and direct coal liquefaction, are under consideration. However, those options are not discussed here since they are beyond the scope of this study.

2.3   Clean coal technologies

Conventional coal-fired plants are large contributors to air pollution. The combustion-generated pollutants of particular significance include oxides of sulphur, nitrogen, and carbon as well as fine organic and inorganic particulates (fly ash, dust, etc.). Coal-fired plant pollution is due to the release of the hot flue gas produced from the combustion of coal into the atmosphere. Dust from power plants has been linked to cancers; SO_2 and NO_x have both been identified as acid rain precursors. NO_x has also been associated with the production of photochemical smog and causing danger to people's health. Mercury, being extremely toxic, can cause both acute and chronic poisoning. Various pollutant control systems have been developed over the past several decades and are continually evolving. These new technologies that facilitate the use of coal in an environmentally more friendly way, reducing drastically their pollutant emissions, is what is commonly known as clean coal technologies (CCT). Within this concept two different approaches can be considered:

- The first approach consists of reducing emissions by reducing the formation of pollutants during the combustion process and/or cleaning the flue gases after combustion. Most of these techniques are commercially available, backed by large-scale operating experience, as outlined in Sections 2.3.1, 2.3.2, and 2.3.3.
- The second approach for CCT is to develop systems with higher thermal efficiency so that less coal is consumed per unit of power generated, together with improved techniques for flue gas cleaning and for residue use or disposal. Within this concept, several technologies have appeared; most of them are at a demonstration stage, as outlined in Section 2.3.4.

2.3.1   Particulate control

Devices for the control of particulate emissions have been in commercial operation for the past half-century. Among these systems are cyclones, fabric filters, and electrostatic precipitators. A more recent development has been the development of ceramic candle filters.

In today's coal-fired plant systems, cyclones are typically used in tandem with other particulate control systems because they are not adequate for removal of all particulate sizes. Advantages of cyclones are ease of maintenance (no moving parts) and the ability to withstand harsh conditions. The disadvantage of cyclones is their inability to effectively remove particles below 5 μm in size due to the low angular momentum of these particles.

Fabric filters, also known as bag houses, are constructed of a variety of materials and positioned downstream of the combustion chamber. Advantages of their use include high particulate removal rates, up to 99.99%. Disadvantages include high maintenance costs, since the filters must be cleaned and maintained frequently, and the inability of the filter materials to withstand high-temperature conditions.

Electrostatic precipitators (ESPs) constitute probably the most widely used particulate removal system for utility-scale coal-fired plants. An electrostatic precipitator consists of a spray electrode (usually a fine wire) of negative polarity and a collecting electrode (a plate) of positive polarity, which is grounded. Voltage between these electrodes is usually on the order of several thousand kilovolts DC. As particle-laden flue gases pass by the spray electrode, the particulate matter is ionised and attracted to the collecting electrode plate. Periodically, a mechanism is used to dislodge the collected particles from the surface of the collecting electrode so that the particles can fall to the bottom of the precipitator for collection. An arrangement of multiple spray and collecting electrodes is usually used in practice to handle the high volume of flue gas. This process can take place over a wide range of temperatures, with larger plate surfaces required for high-temperature operation. Electrostatic precipitator removal efficiencies can reach nearly 100% for particles down to 1 μm in size, with decreasing effectiveness for smaller particle sizes.

Ceramic or metallic candle filters operate along the same principle as fabric filters. However, these filters are designed to withstand higher-temperature and increased pressure flue gas conditions, since they are constructed from sturdier materials. The cooling of flue gas for treatment results in a measurable loss in net plant efficiency, even when the heat from the flue gas is recuperated and used in the process. In the cases of PFBC and IGCC technology, there is a great need to develop systems for the cleaning of gas at increasingly higher temperatures and pressures. These factors should lead to increased development of a ceramic candle filter technology, which has not yet reached full maturity.

The level of particulate control is affected by coal type, sulphur content, and ash properties. Greater particulate control is possible with enhanced performance units or with the addition of wet ESP after FGD.

2.3.2 SO_2 control

SO_2 is created during the combustion process when trace amounts of sulphur present in the coal are oxidised by combustion air. Flue gas desulpherisation (FGD) can be accomplished by dry injection of limestone into the ductwork just behind the air pre-heater (50–70% removal) with recovery of the solids in the ESP. For fluidised bed combustion units, the fluidised bed consists primarily of limestone, which directly captures most of the SO_2 formed. For PC units, wet flue gas desulphurisation (wet lime scrubbing) can achieve 95% SO_2 removal without additives and more than 99% SO_2 removal with additives [17]. The wet limestone scrubber FGD system is typically located along the flue gas ductwork, downstream of the particulate control device. The theory of operation of a wet limestone scrubber FGD system relies on the chemical reaction amongst limestone, sulfur dioxide, water, and oxygen to produce gypsum, water, and carbon dioxide. This reaction occurs as follows:

$$CaCO_3 + SO_2 + 2H_2O + \frac{1}{2}O_2 \Rightarrow CaSO_4 + 2H_2O + CO_2 \tag{2.1}$$

In this system, the flue gas is directed into a large vessel, where nozzles mounted near the top spray a fine mist of limestone powder/water suspension. The flue gas enters near the bottom of the vessel and exits at the top, so it passes directly through the spray. As the misted suspension falls towards the bottom of the vessel, it absorbs SO_2 in the flue gas and is collected in the bottom of the vessel, which is filled with a pool of the same limestone/water suspension. An air inlet near the bottom of the tank provides oxygen to complete the reaction. Thus, small particles of gypsum are formed in the pool. As these particles fall to the bottom of the pool, they are filtered out, and subsequent processes produce a gypsum by-product. Meanwhile, the de-sulphurised flue gas is exhausted and reheated in a heat exchanger by fresh flue gas entering the vessel, such that the de-sulphurised flue gas can flow out the plant's stack.

Coal power plants generate approximately 65% of the human-generated mercury emissions into atmosphere [18]. Today, about 25% of the mercury in the burned coal is removed by the existing flue gas treatment technologies in place, such as ESP or fabric filters. Wet FGD can achieve 60% mercury removal. In combination with SCR, mercury removal could approach 95% for bituminous coals [19]. For subbituminous coal and lignite, however, mercury removal is typically less than 40%, even when the previously outlined flue gas cleaning technologies are applied. An improvement could be achieved through application of activated carbon or brominated activated carbon injection. The control of mercury emissions from coal combustion remains a serious issue, and mercury removal is expected to reach 90% in coming decades.

2.3.3 NO_x control

Nitrogen oxide (NO) and nitrogen dioxide (NO_2), known collectively as NO_x, are formed during the combustion of coal by one or more of the following three chemical mechanisms: (1) "thermal" NO_x, resulting from oxidation of molecular nitrogen in the combustion air; (2) "fuel" NO_x, resulting from oxidation of chemically bound nitrogen in the fuel, and (3) "prompt" NO_x, resulting from reaction between molecular nitrogen and hydrocarbon radicals.

Thermal NO_x typically represents up to 20% of the NO_x emitted during pulverised coal combustion in utility boilers. Its rate of formation is directly proportional to the exponential of temperature and to the square root of the oxygen concentration. At temperatures above 1540 °C, significant amounts of thermal-NO_x are produced through dissociation and oxidation of molecular nitrogen from combustion air. Therefore, control is accomplished by moderating flame temperature and oxygen concentration. For PC boilers, practical methods for thermal-NO_x control include the following measures: (1) increasing the size of the combustion zone for a given thermal input by using an overfire air (OFA) system in case of an existing boiler or by increasing the furnace dimensions in the burner area in case of a new boiler; and (2) reducing the rate of combustion and, consequently, lowering peak flame temperatures with specially designed burners, so-called low-NO_x burners.

In pulverised coal combustion in a utility boiler, fuel NO_x may typically contribute up to 80% of the NO_x emissions. Formation of fuel NO_x depends on the nitrogen content in the fuel and on the amount of oxygen available to react with the nitrogen during coal devolatilisation in the early stages of combustion. Measures for reducing oxygen availability include lowering the excess air level and/or controlling the rate at which fuel and air mix (e.g., staged combustion) such that an initial fuel-rich zone is followed by a burnout zone. However, these techniques tend to increase unburned combustibles (CO and/or carbon particulates). More unburned combustibles occur even if the total air introduced to the process is the same. This can be due to incomplete mixing of air and fuel or due to insufficient residence time to complete combustion after the air is completely introduced. Prompt NO_x contributes a relatively minor fraction of total NO_x emissions for coal-fired boilers.

Without emission controls, NO_x may exceed 1000 ppm from a boiler, or 0.1% of the outlet gases. NO_x formation can be limited by modifications to the combustion process and by changing fuel. Once formed, NO_x can be reduced by injecting reagents into the furnace or in a subsequent reactor that includes solid catalysts, selective catalytic reduction (SCR) systems, downstream of the boiler. SCR is required to accomplish the lowest levels of NO_x. However, it is usually more cost effective to limit NO_x as much as possible within the combustion system, then achieve the remaining reduction by SCR. Combined use of combustion modifications and SCR can reduce NO_x to less than 50 ppm (0.005%). Examination of the chemical paths of nitrogen oxides formation and destruction in flames led to the formulation of guidelines for primary measures of NO_x emissions reduction in boilers:

- Reducing the peak flame temperature by heat extraction and/or by flue gas recirculation
- Diluting the reactant concentrations by flue gas or steam mixed with gaseous fuels and recirculated burned gas mixed with combustion air
- Staging the combustion air to produce fuel-rich/fuel-lean sequencing favourable for the conversion of fuel bound nitrogen to N_2
- Staging the fuel so that the NO formed earlier in the flame is getting reduced by its reactions with hydrocarbon radicals ("NO reburning")

The optimum control system may contain one or several of these measures, selected depending on the capacity of the unit, fuels to be fired, and NO_x reduction requirements.

2.3.4 Energy efficiency

Supercritical and ultra-supercritical conditions

Raising the steam pressure and temperature to 24 MPa/550 °C at the turbine inlet increases plant efficiencies to the 43–45% (LHV) range [10], what is a significant progress over traditional subcritical units. Although some of the earlier supercritical units experienced various problems related to operation and reliability, such as long startup times, slagging, cracking of water wall tubes, high air leakage due to pressurised furnace, and in consequence low availability, today a large number of supercritical units have been put into operation and are achieving a good operational performance. Super-critical units are now available in sizes over 1 GW_{el} for PF boilers and around 0,5 GW_{el} for CFB boilers (expecting a scale-up to 0,8 GW_{el}).

The development of improved high-strength ferritic and austenitic materials has made possible the way for ultra-supercritical units operating at even higher temperatures ($\geq 600\,°C$) and pressures (≥ 30 MPa), with potential single-cycle efficiencies around 50%. Examples of these improvements have been demonstrated in Germany, Denmark, and Japan. Future research and developments attempt to a new generation of ultra-supercritical units reaching steam parameters near 40 MPa and 750 °C.

Pressurised fluidised bed combustion

Pressurised fluidised bed combustion (PFBC) integrates a combined cycle with sorbent injection for SO_2 reduction and particulates removal from flue gases. In this process pressurised combustion systems produce, as a result of the oxidation reactions in the reactor, a hot combustion gas that drives a gas turbine and produces steam by heat transfer to water tubes immersed in the bed that moves a steam turbine. Typically 80% of the power comes from the steam turbine. Pressurised fluidised bed boilers operate at pressures of 1–1.5 MPa with combustion temperatures of 800 °C. The injection of a sorbent in the bed reduces SO_2 emissions, making later flue gas desulphuration systems unnecessary. On the other hand, particulate gas cleaning systems are critical in the plant to prevent residual solids from entering the gas turbine. Although, as with atmospheric FBC, it is possible to use bubbling or circulating beds, currently commercial-scale operating units are of the bubbling bed type. Compared to FBC, the heat release rate per unit bed area in PFBC is several times higher and the bed height is 3–4 m instead of the typical bed height of 1 m in FBC. This yields a larger carbon inventory in the fluidised bed and lower NO-emissions due to the reduction of the NO formed by solid carbon in the bed. However, the high carbon load does not reduce the emission of N_2O (typically in the range of 50–100 ppm), which is still stable at the relatively low temperatures of the PFBC [10]. The low temperatures also are unfavourable for an efficient usage of the gas turbine.

Nowadays, although PFBC units suffered certain operational problems, larger units of 360 and 250 MW_{el} have been started up in Japan. PFBC units reach efficiencies of around 40%, and with the development of more advanced cycles they are intended to achieve efficiencies over 45%. Future developments, called advanced PFBC systems, aim to increase the temperature at the inlet of the gas turbine from approximately 850 °C to approximately 1350 °C, attaining a more efficient power generation (approximately 46–47% net efficiency) through combining PFBC and fluidised-bed gasification technologies integrated in the same plant [10].

Integrated gasification combined cycle

An integrated gasification combined cycle (IGCC) comprises gasification of coal (using different types of gasifiers), syngas stream cleaning systems for H_2S, chlorides and particulates, and subsequent combustion in a gas turbine. IGGC uses a combined cycle with a gas turbine driven by the combusted syngas and a steam turbine driven by the steam produced by the gasifier and by heat recovered from exhaust gases leaving the gas turbine. Typically 60–70% of the power comes from the gas turbine. Coal gasification takes place under an air/oxygen shortage in a pressurised reactor, and the product gas, a mixture of mainly CO and H_2 called syngas, is cleaned and then burned, generating high-temperature and high-pressure combustion products. IGCC is the cleanest advanced coal technology available today. It is also demonstrated to be working with no major operational problems. The future of IGCC depends on whether it will be possible to reduce its investment costs and to increase the cycle efficiency. The high costs are related to the required oxygen plant and to lower integration level of various subsystems, such as the air separation unit, the syngas cooler, the clean-up, and the gas turbine. Efficiencies reported for large-scale coal-fired IGCC plants ($\sim$0.3 GW_{el}) can reach 43–44%, a remarkable figure considering the low temperature of the gas cleaning stage, which restricts the attainable overall efficiency. Recently the IGCC technology has gained a new perspective as it provides favourable conditions for CO_2 sequestration and possibilities for efficiencies exceeding 50% by the production of hydrogen and combination of IGCC with fuel cell technology.

2.4 Carbon capture and storage technologies

The global mean atmospheric carbon dioxide (CO_2) concentration measured over the ocean surface has increased from about 315 ppm in 1958 to 395 ppm at the beginning of 2013 [20] (Figure 2.3), thus raising serious concerns about its potential impact on climate change. Coal combustion for electric power generation is one of the major

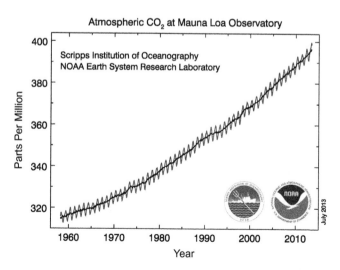

Figure 2.3 Atmospheric mean CO_2 measured at mauna loa observatory, Hawaii, 1960–2010. The red line represents the monthly mean values. The black curve represents the same, after correction for the average seasonal cycle [20].

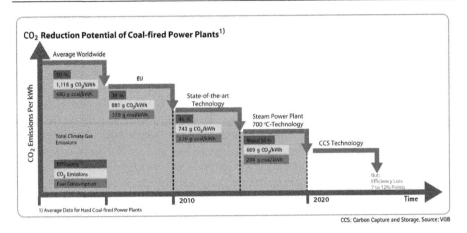

Figure 2.4 CO_2 reduction potential in coal-fired power plants, 2000–2020 and beyond.

contributors to anthropogenic CO_2 emissions to the atmosphere. Therefore, the development of an emission-free, coal-fired power plant would be an important step toward the reduction of human-made greenhouse gas emissions into the atmosphere.

There are two primary ways of reducing CO_2 emissions from coal use:

- Improving efficiencies at coal-fired power stations, leading to lower emissions per unit of energy output, as discussed in Section 2.3.4; and/or
- Applying carbon capture and storage (CCS) technologies, which can reduce CO_2 emissions to the atmosphere by 80–90%

An increase in the efficiency of the older power plants to a state-of-art technology level can reduce the emissions worldwide, with 33% from 1,116 down to 743 $g_{CO_2}/$ KWh, as shown in Figure 2.4. Successful development and implementation of the 700 °C technology can further reduce the emissions to 670 $g_{CO_2}/$kWh. However, the greatest potential for reduction of CO_2 emissions is offered by CCS technologies.

Some CCS methods aim to clean the exhaust gases after combustion by using absorption and/or adsorption (post-combustion measures including flue gas washing with chemicals). Other methods extract carbon from the fuel (pre-combustion measures, mainly IGCC) or use indirect combustion processes (chemical looping). All of these strategies require large mass exchangers. Some of these techniques produce impure streams of CO_2 and do not remove all of the carbon dioxide from the exhaust gas. According to Yantovsky et al. [21], the only form of combustion with truly zero CO_2 emissions[5] in existence today is pre-combustion gas separation, namely, the combustion of fuel using oxygen instead of air. Although it is theoretically possible to burn pulverised fuel in pure oxygen, this would increase flame temperatures, thus also increasing NO_x and ash slagging as well as fouling problems. Therefore, to moderate the flame

[5]Here zero CO_2 emissions means that the CO_2 produced during combustion is simply not released into the atmosphere.

temperature and to reduce NO_x and ash-related issues, the oxygen is mixed with recycled flue gas (RFG). The exact percentage of oxygen depends on whether the RFG is dry or wet. In either case, to avoid damaging the RFG fans through erosion, the RFG extraction point occurs after the particulate cleanup stage. This approach to reducing CO_2 emissions from power plants is often called oxy-firing or oxyfuel combustion and is depicted schematically in Figure 2.5. The resulting flue gas consists primarily of CO_2 and water vapour along with some N_2, O_2, and trace gases such as SO_2 and NO_x. The flue gas can be processed relatively easily (e.g., through rectification or distillation) to enrich the CO_2 content in the product gas further to values of 96% to 99% [19]. The purified CO_2 stream is then compressed and condensed to produce a manageable effluent of liquid CO_2, which can be sequestered for storage (CCS) or for use in subsequent processes (CCR).

To provide the oxygen necessary for an oxyfuel process, two main technologies are currently developed: oxygen supply by a cryogenic technique (Figure 2.5) or by using a high-temperature ceramic ion-transport membrane (ITM) (Figure 2.6).

The first method is a mature technology that is already well established in industry and therefore can be easily implemented. The energy demand for oxygen production, however, is about 240 kWh per ton O_2, which, together with CO_2 compression, results in a net efficiency drop at a power plant of around 8–10% points [22–23]. This energy penalty can be further reduced by optimisation of the cryogenic air separation, whereby

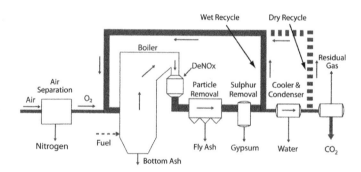

Figure 2.5 Oxyfuel combustion scheme with wet and dry recycle of the flue gas.

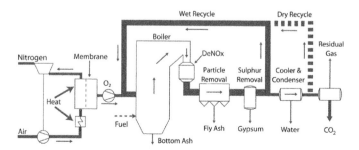

Figure 2.6 Oxyfuel combustion scheme with ITM for oxygen supply (without heat integration).

values around 160 kWh per ton O_2 seem to be the optimistic technological limit [24]. The second method for oxygen supply is based on a high-temperature oxygen ion transport membrane (ITM) that, due to the reduced auxiliary power required for oxygen production, appears to be the more cost- and energy-effective alternative to the cryogenic process [25,26].

The development of ITM is concentrated on mixed-type (ionic and electronic) conductor materials, typically with perovskite or fluorite molecular structures. Because these membranes are dense, impermeable ceramics, no gas can cross this barrier, and so the separation efficiency will be 100% for oxygen. The permeability of such materials is strongly dependent on temperature, thickness, and oxygen partial pressure difference between the two sides of the membrane. Therefore, heat integration of the ITM into the process can lead to significant reduction in operational costs.

The integration of ITM into the oxyfuel process, however, requires involvement of new components, thus modifying the whole power plant process. The OXYCOAL-AC process [27] shown in Figure 2.7, bottom, for example, considers that the membrane will be heated and kept at its operating temperature by recirculation of flue gas that enters the membrane module at temperatures of around 850 °C. Since the ceramic membrane is susceptible to dust and sulphur, the flue gas has to be cleaned in a hot gas filtration upstream of the membrane. To capture sulphur together with the dust,

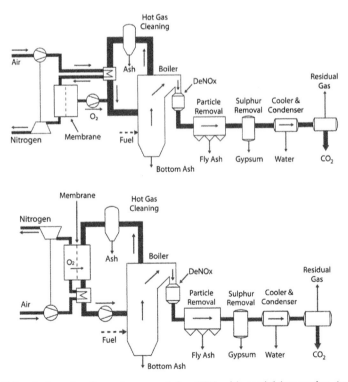

Figure 2.7 Oxyfuel combustion scheme applying ITM with partial integration (top) and full integration (bottom) into the process.

DeSO$_x$ measures have to be implemented, e.g., adding limestone before the filtration or directly into the furnace. The hot gas filtration unit consists of ceramic filter candles and is also operated in the temperature range around 850 °C. The filter candles are cleaned by gas pulses that blow off the filter cake. Thus, the "cleaned" flue gas enters the membrane module, where it is enriched with oxygen. Besides the high temperature of an ITM, a partial pressure difference of oxygen across the membrane has to be established. Oxygen is then transported from the high-pressure, oxygen-rich side (feed side) through the membrane into the low-pressure, oxygen-lean side (sweep side). To maintain the partial pressure ratio of oxygen, the feed side is pressurised with air and the sweep side is swept with flue gas, which is thereby enriched with oxygen. Before entering the membrane module, the pressurised air has to be pre-heated in the air-RFG heat exchanger to a temperature of about 750 °C in order not to cool down the membrane. The mean temperature of the membrane under these conditions is about 825 °C. Finally, the recirculation is closed by a RFG fan, which drives the oxygen-enriched flue gas back to the burners. The hot nitrogen is expanded in a turbine that is used to drive the air compressor.

Applying this technology on a large-scale, state-of-the art power plant burning bituminous coal can reduce the efficiency drop to around 6.6 points compared to the reference air operation [28]. However, membrane materials having the highest permeation rate at the conditions of interest are not resistant to CO_2 and SO$_x$. Therefore, the development of new membrane materials that do not lose their permeation abilities in direct contact with flue gas is an ongoing issue [29]. Since such materials are not yet available, an alternative solution is under consideration (see Figure 2.7, top) that avoids direct contact between flue gas and the membrane. The oxygen, in this case, is extracted from the membrane module by a vacuum pump.

Hence, the development of the core technology and the system integration of an oxy-firing process that is based on membrane oxygen separation becomes a key issue on the way to reducing significantly the energy penalty related to all CCS technologies. Thus the oxyfuel technology will become commercially attractive, which is very important if carbon capture is going to become mandatory in a future regulatory framework.

Other critical issues that require intensive R&D before applying the oxycoal technology include these:

- In terms of recirculation of flue gas: The determination of the "place of extraction" is a function of the type of RFG, namely, cold and dry, cold and wet, or hot and wet RFG
- In terms of corrosion: Material choice, controlled flame temperatures, and excess oxygen ratio
- In terms of combustion stability: Fame temperature, flame shape, PF and oxygen concentrations, partial loads, flame stability, the volume of the RFG, and the place of FG extraction
- In terms of startup and shutdown conditions: Burner design and burner settings
- In terms of thermal efficiency: Heat transfer inside the furnace and in the convective part
- In terms of ASU and CO_2 compression: Oxygen quality, FG composition, emissions, etc.

- In terms of slagging and fouling: Ash quality, combustion gas composition, and flame temperature
- In terms of burner design: Fame characteristics, swirl ratios, oxygen concentration and oxygen distribution among the different streams, gas velocities, number of burners, etc.
- In terms of boiler design: Retrofit or new designs.

Finally, it can be concluded that oxycoal is a zero-CO_2-emission emerging technology with strong commercial interest. Combustion science and modelling are needed to advance the oxy-firing technology and to optimise operations, particularly when this effort is linked to pilot-plant trials and to further design of plants. There are already several demonstration projects planned for PF oxyfuel technology, a 30 MWth pilot plant and a 250 MW demonstration plant in Schwarze Pumpe, Germany; a 30 MWth retrofitted boiler in Biloela, Australia, and a 30 MWth pilot plant and a 323 MW full-scale demonstration plant in El Bierzo, Spain. The realisation of these projects will bring valuable information that is necessary to understand the effect of a CO_2-rich atmosphere on the oxy-combustion of PF in real scales and important know-how for the potential scale-up and construction of a CO_2 emission-free coal-fired power plant.

2.5   Summary of Chapter 2

A recent study on the primary energy demand and on the reserves of fossil fuels worldwide showed that coal has the biggest proven reserves among the fossil fuels and it is uniformly distributed to the world, thus predefining its important role in the energy mix of the future.

The existing state-of-the-art coal combustion technologies are reviewed with special attention given to clean coal technologies leading to significant reduction of the emissions from pulverised coal firing.

Carbon dioxide capture and storage (CCS), together with improved energy efficiency technologies, are identified as critical technologies enabling significant reduction of CO_2 emissions while allowing coal to meet the present world energy needs.

Oxyfuel pulverised coal combustion appears to offer significant potential for new plants or retrofit CO_2 capture applications. Basic research to develop less costly oxygen separation technologies is a high priority in order to minimise the efficiency loss in the overall process. Different concepts of process integration, based on ion transport membranes, were evaluated. Much research is needed on the compositional requirements for the boiler, on process design and evaluation studies, and on process development units.

Replacing N_2 with recycled flue gas for dilution purposes in oxy-firing leads to significant changes in the gas properties inside the combustion chamber. Carbon dioxide exhibits pronounced differences in thermodynamic and optical properties compared to air. Hence, it can be concluded that it is of vital importance for an efficient management of the combustion process at first to understand how the particular properties of CO_2 affect the chemical reactions and the heat transfer that take place during PF oxycombustion.

Part I

Theoretical Aspects

3 Theoretical Aspects of Burning Pulverised Fuel in CO_2 Atmosphere

The combustion of pulverised coal is a complex process governed by a number of physical and chemical phenomena. The principal steps of the reaction progress are the thermal decomposition of the raw coal and the subsequent burnout of the char and the volatile matter. The following reaction steps (3.1– 3.6) typically summarise the main process of coal combustion:

$$\text{Coal (raw)} \rightarrow \text{Char} + \text{Volatiles} + H_2O \tag{3.1}$$

$$\text{Volatiles} + O_2 \rightarrow CO + H_2O + SO_2 + NO \tag{3.2}$$

$$CO + \frac{1}{2}O_2 \rightarrow CO_2 \tag{3.3}$$

$$\text{Char} + \frac{1}{\phi}O_2 \rightarrow \left(2 - \frac{2}{\phi}\right)CO + \left(\frac{2}{\phi} - 1\right)CO_2 \tag{3.4}$$

$$\text{Char} + CO_2 \rightarrow CO + CO \tag{3.5}$$

$$\text{Char} + H_2O \rightarrow H_2 + CO \tag{3.6}$$

As it is heated, a coal particle undergoes decomposition into char and volatile material (reaction 3.1), the former burning slowly in the later stages of the flame (reactions 3.4–3.6), whilst the volatile material is assumed to rapidly form CO (reaction 3.2) and subsequently CO_2 (reaction 3.3) as the most simple reaction mechanism. In the case of combustion of PF in an O_2/RFG mixture, the high partial pressure of CO_2 at the surface of hot burning particles results in higher concentrations of surface complex C(O) on the carbon surface, which triggers the reaction of CO_2 with the char carbon to form CO. The CO diffuses away from the surface through the boundary layer, where it combines with the inward-diffusing O_2 according to reaction 3.3. Of course, many elementary reaction steps are involved in reaction 3.3, with one of the most important being $CO + OH \Leftrightarrow CO_2 + H$, as discussed in Chapter 5. In case of oxycombustion with wet recycling, the H_2O gasification reaction 3.6 can play a significant role in char burnout as well.

Replacing N_2 with recycled flue gas (containing mainly CO_2) for dilution purposes in oxy-firing will change the gas properties in the combustion chamber. Therefore, the effect of the changed gas properties on the homogeneous and heterogeneous reactions

Combustion of Pulverised Coal in a Mixture of Oxygen and Recycled Flue Gas. http://dx.doi.org/10.1016/B978-0-08-099998-2.00003-5

as well as on the heat transfer that takes place during PF oxycombustion is discussed in the present chapter with respect to fundamental coal combustion theory.

3.1   Differences between air and oxyfuel combustion

Conventional PF coal-fired boilers, currently being used in the power industry, use air for combustion in which the nitrogen from the air (approximately 79% vol) dilutes combustion products such as CO_2 and water vapor in the flue gas. During oxyfuel combustion, a combination of oxygen (typically of greater than 95% purity) and recycled flue gas is used for combustion of the fuel. A gas consisting mainly of CO_2 and water vapour is generated, with a concentration of CO_2 between 70% and 90%, depending on recycle mode (wet or dry). The recycled flue gas is used to control flame temperature and make up the volume of the missing N_2 to ensure that there is enough gas to carry the heat through the boiler.

Oxyfuel combustion performed in bench- [30], pilot- [31], and demonstration-scale [32] experiments has been found to differ from air combustion in several ways, including reduced flame temperature, delayed flame ignition and related flame instability, reduced NO_x and SO_x emissions, and changed heat transfer. Many of these effects can be explained by differences in gas properties between CO_2 and N_2, the main diluents in oxyfuel atmosphere and air, respectively. Carbon dioxide has different thermo-physical and optical properties than nitrogen that influence both combustion reaction rates and heat transfer.

Table 3.1 shows some selected gas properties of N_2 and CO_2. The molar heat capacity of CO_2 is higher than that of N_2, with CO_2 thus being a bigger heat sink than nitrogen. Hence, in order to maintain the adiabatic flame temperature (AFT) at the same level as for air combustion, an increase of the O_2 vol% in the O_2/RFG mixture is required.

The molecular weight of CO_2 is 44 g/mole, compared to 28 g/mole for N_2. Thus the density of the flue gas is higher in oxyfuel combustion. This results in lower gas velocities and higher residence times of particles in the furnace. The oxygen diffusion rate in CO_2 is 0.8 times that in N_2, thus affecting the oxygen availability at the char surface. The lower thermal diffusivity of CO_2 leads to a slower flame propagation speed. The higher energy per volume for CO_2 results in a lower temperature of the

Table 3.1 Gas properties for N_2 and CO_2 at 900 °C [33].

Property	N_2	CO_2	Ratio CO_2/N_2
Thermal conductivity, 10^{-3}, W/mK	74.67	81.69	1.09
Molar heat capacity c_p, kJ/kmol K	33.6	56.1	1.67
Density ρ, kg/m^3	0.29	0.45	1.55
O_2 diffusion coeff. Do, 10^{-4}, m^2/s	3.074	2.373	0.77
Thermal diffusivity, 10^{-7}, m^2/s	2168	1420	0.65
Molecular weight M, kg/kmol	28	44	1.57
Energy per volume, $\rho c_p M^{-1}$J/m^{3}K	0.34	0.57	1.67

combustion gases compared to air-firing (if O_2 content is kept at the same level as for air). The high partial pressures of CO_2 and water vapour in the RFG result in higher flue gas emittance. Thus, similar radiative heat transfer will be attained for a boiler retrofitted to oxyfuel when the O_2 content in the O_2/RFG mixture passing through the burner should be less than the required levels for the same AFT.

The effect of the changed gas properties on the homogeneous and heterogeneous reactions that take place during PF oxyfuel combustion as well as on the related heat transfer in the boiler is a subject of detailed investigation, as discussed in more detail in Sections 3.2, 3.3, and 3.6, respectively.

3.2   Coal devolatilisation and particle ignition

Devolatilisation behaviour is the most important aspect of coal quality in combustion chemistry, being primarily responsible for the partitioning between volatile matter and char. Thus it determines: (1) particle residence time requirements; (2) near-burner heating rates that govern ignition, flame stability, and flammability limits of pulverised coal flames; and (3) soot loadings that determine near-burner radiation fluxes. Volatiles released during devolatilisation can account for up to 50% of the heating value of the coal [34]. Ultimate yields have been shown to be very similar at about 50 wt% for coals through the high-volatile bituminous coal and lignites, then diminish in coals of higher rank. However, the proportions of gases and tars vary widely, with gas dominating the yields of low-ranked coals. The mechanisms and variables that control coal devolatilisation are discussed in detail by Smoot [15] and Smith *et al.* [35]. Only a brief description of coal devolatilisation is given here, emphasising on small coal particles (typical for PF firing) where devolatilisation is usually kinetically controlled, with no internal particle temperature gradients. Special emphasis is given to the most recent investigations on the devolatilisation behaviour of coal particles in oxyfuel atmosphere.

3.2.1   Devolatilisation mechanisms and features

Devolatilisation (or pyrolysis) is the first stage in coal combustion. Devolatilisation occurs as the raw coal is heated in an inert or oxidising atmosphere. The devolatilisation stage consists of three distinct physical processes: (1) pyrolysis (the decomposition chemistry), (2) volatile transport through the pores, and (3) secondary reactions that can change the chemical products of gas and/or cause decomposition of volatile products on the walls of the pores.

The pyrolysis behaviour of coal is affected by *temperature, heating rate, pressure, particle size,* and *coal type,* among other variables [15,34]. Higher mass release during devolatilisation generally occurs at higher temperatures.

As the *temperature* of the coal increases, the bridges linking the aromatic clusters break, resulting in finite-size fragments that are detached from the macromolecule [35]. The bridges consist of a distribution of different types of functional groups, and the weakest bonds are broken first. The fragments are commonly referred to as metaplast or liquid coal components. This fluidity normally occurs in coals with 81–92 wt% carbon, but it depends on oxygen and hydrogen content in the coal as well as on the heating rate. Plastic properties become more pronounced at high heating rates, up to a point; if the heating rates become too high, coals cannot plasticise or fluidise, because cross-linking reaction temperatures are rapidly reached before the coal structure can relax and fluidise.

Further, the metaplast either (1) vaporises and escapes the coal particle, or (2) crosslinks back into the macromolecular structure. The metaplast that vaporises, referred to as tar, is a mixture of aromatic compounds with average molecular weights in the range of 350 whose chemical structure closely resembles that of the parent coal [36]. The tar is the primary source of soot, which dominates radiative heat transfer in the volatile flame region.

Side chains and the broken bridge material are released as light gas in the form of light hydrocarbons and oxides.

As coal pores melt and fuse, the subsequent formation of bubbles (filled with light gases and tar vapour) results in swelling. Depending on heating rate, temperature, and particle size, either a particle may swell or bubbles may rupture. In PF firing, the bubbles normally rupture due to high heating rates, but significant changes occur in the char morphology (porosity and internal surface area) due to bubble formation [15]. Thus, the particle softening affects the ignition, particle trajectory in the furnace, reactivity, and eventual fragmentation and size distribution [36].

In case of *pressurised conditions*, the more volatile components of tar are held in the particles for longer times, decreasing viscosity at the critical time of gas evolution. With a further increase in pressure, the compressive external environment results in a reduced swelling [37,38].

Volatiles are transported via formation of bubbles. Conversely, high-rank coals usu-ally exhibit little fluidity and plasticity, and retain their pore structure during devolatil-isation. In this case, volatiles are transported by diffusion via pore structures.

The *heating rate* has the following two effects on devolatilisation behaviour: (1) as heating rate increases, the temperature at which volatiles are released increases; and (2) generally, as heating rate increases, the overall volatiles yield increases [15,34]. Heating rate effects can be explained by the competition between tar formation (bridge breaking), tar destruction (cross-linking), and tar evolution (mass transfer), all of which depend differently on temperature.

Particle size effects become small at diameters below 200 μm and are usually ignored in pulverised coal applications.

Devolatilisation behaviour is largely dependent on *coal type* [39]. Low-rank coals (lignites and subbituminous coals) release a relatively large amount of light gases and less tar. Bituminous coals release much more tar than lower-rank coals and moderate amounts of light gas. The highest-rank coals release only small amounts of tar and even lower amounts of light gas.

Light gas released during devolatilisation consists mainly of methane, carbon diox-ide, carbon monoxide, and water vapour [34,40]. Other constituents include low molec-ular weight hydrocarbons such as olefins, nitrogen species, and sulfur species. Saxena [34] studied light gas release at atmospheric pressures and low heating rates (in the order of 1 K/sec). Methane and small amounts of olefins began to evolve at temperatures between 473 K and 773 K, whereas nitrogen structures and organic sulfur species began to decompose. Hydrogen began to evolve around 673 K. At higher temperatures (773–973 K), the volume of hydrogen, carbon dioxide, and methane increased relative to other hydrocarbon species.

Suuberg et al. [41] studied the devolatilisation behaviour of a lignite at a heating rate of 1000 K/s. Carbon dioxide evolution was observed to begin at about 723 K. Chemically formed water and carbon dioxide evolved in the range of 773–973 K. Between 973 K and 1173 K, hydrogen and hydrocarbon gases were released. At higher temperatures, the formation of additional carbon oxides was observed.

The composition of the light gas released during devolatilisation is a function of coal rank [42]. Light gas released from lignites contains a relatively large amount of carbon dioxide and carbon monoxide but contains only a small amount of methane. Light gas evolved from bituminous coals during devolatilisation contains a smaller fraction of carbon dioxide and carbon monoxide and a larger fraction of methane compared to light gas evolved from lignites. The variations in the species distribution of light gas as a function of rank are believed to be the result of variation in the composition of the aliphatic side chains.

3.2.2 Coal particles ignition

Investigations of ignition mechanisms showed that the ignition of char particles typically occurs heterogeneously, whereas coal particles may ignite heterogeneously (caused by the coke-oxidation reaction), homogeneously (caused by volatile-matter oxidation in the gas phase), or by a combination of both mechanisms [43–45]. Homogeneous ignition is favoured by high oxygen concentrations and close particle interactions. Ponzio et al. [44] have investigated the ignition behaviour of coal particles at different oxidiser temperatures (870–1273 K) and oxygen concentrations (5–100 mole%). The ignition behaviour was classified into (1) sparking ignition by heterogeneous oxidation of the nondevolatilesed coal at high-oxidiser temperatures and medium- to high-oxygen concentrations, (2) flaming ignition by homogeneous ignition of volatiles for high-oxidiser temperature and low-oxygen concentrations, and (3) glowing surface ignition by heterogenous oxidation of char at low-oxidiser temperature and low-oxygen concentrations. It was shown that the ignition time decreases with increasing oxidiser temperature and increasing oxygen concentrations. Furthermore, it was indicated that the influence of oxidiser temperature and oxygen concentration was smallest at high-oxidiser temperature. Baum and Street [46] reported that heterogeneous ignition occurs not only on the outer surfaces of particles but also inside the pores. They concluded that the ignition and the combustion of the great majority of particles within size range typical for PF firing is chemically controlled. At high surface oxidation temperature, two ignition jumps, the first due to the heterogeneous mechanism and the second due to the homogeneous mechanism, were observed by Gururajan et al. [47]. Consideration of the simultaneous occurrence of volatile combustion and surface oxidation shows that surface oxidation influences the ignition temperature of only small particles or at high oxygen concentration. Model predictions by Mitchel et al. [48] showed that relatively little CO_2 is formed in the boundary layers of small particles (less than $100 \, \mu m$). Correspondingly, little thermal energy is transferred to the particle surface as a result of CO conversion in the boundary layer. Thus, for small particles, any CO_2 formation must occur on the particle surface, not in the surrounding gas.

3.2.3   Effects of CO_2 on coal devolatilisation and on particle ignition

The presence of high concentrations of CO_2 in the bulk gas could influence the *coal pyrolysis* process in two main ways [49]:

- CO_2 is a product of coal pyrolysis, which may affect the volatile composition and yield.
- CO_2 is a reactant in the char gasification reaction, which may cause differences in the formation of SO_2/NO_x precursors.

Messenboeck *et al.* [50] studied flash pyrolysis of a bituminous coal under three different atmospheres (He, H_2O, and CO_2) at 1 MPa and at a peak temperature, heating rate, and holding time of $1000 \, ^\circ C$, $1000 \, ^\circ C \, s^{-1}$, and 0–60 s, respectively. There were no significant effects of changing the atmosphere from He to CO_2 on the volatile yield until reaching the peak temperature. But afterwards, the reactive gases caused gasification rates much greater than expected from previous reports on char gasification. Simultaneous occurrence of thermal cracking and CO_2 gasification of the nascent char has thus been evidenced, whereas the rates of the gasification seem to be much higher than those reported previously. This may be explained with the findings from Liu *et al.* [51] that the gasification rate of char is very different from that of a raw coal particle (direct gasification without pyrolysis in advance), suggesting an influence of pyrolysis time and atmosphere on char reactivity. This effect was more remarkable for coals with high volatile matter content. Jamil *et al.* [52] have investigated pyrolysis of raw Victorian brown coal particles under atmospheric flow of He and CO_2 in a wire-mesh reactor over wide ranges of heating rates. It was reported that the change of the atmosphere (from He to CO_2) did not have any impact on the yield and composition of released tar. The tar formation occurs so rapidly, even at low temperatures, that it prevents CO_2 from participating in intraparticle reactions. For high heating rates, however, the average rate of gasification was found to be over 20 wt%-daf-coal s^{-1}. It seems that the rate of such rapid CO_2 gasification strongly depends on the rate of thermal cracking. Duan *et al.* [49] reported that during pyrolysis in N_2 atmosphere at $760 \, ^\circ C$, there is a peak in CO_2 release caused by the decomposition of calcite in the coal sample. In CO_2 atmosphere, however, the high partial pressure of CO_2 will prevent the calcite from decomposing and thus will affect the composition of the volatiles. Additionally, in CO_2 atmosphere, the volatiles yield increases as the temperature increases and decreases as the heating rate increases. The presence of CO_2 retards *single coal particle ignition* [53]. According to Yamamoto *et al.* [54], the ignition delay in PF oxy-firing increases, mainly due to the high value of the heat capacity of a gas mixture with higher CO_2 concentration. The results obtained by Suda *et al.* [55], who investigated a PF flame in a small spherical chamber, show that the flame propagation velocity of a PF cloud in a CO_2/O_2 atmosphere decreases to around 1/5 to 1/3 of that in a N_2/O_2 atmosphere at the same O_2 concentration level. This was found to be mainly due to the larger heat capacity of CO_2. An acceleration of particle ignition for both N_2/O_2 and CO_2/O_2 atmospheres was observed when increasing the oxygen concentration. Shaddix and Molina [53] also reported that particle devolatilisation proceeds more rapidly with higher O_2 concentrations and decreases with the use of CO_2 diluent because of the influence of these two species on the mass diffusion rates of O_2 and fuel volatiles. Therefore, an increased oxygen concentration for PF oxy-firing, if selected correctly, can in principle produce ignition times and volatile flames similar to those obtained under PF–air combustion conditions [53]. Molina *et al.* [56] studied the ignition of groups of particles of high-volatile bituminous coal in an optical entrained flow reactor (at Sandia National Laboratories) under oxygen concentrations ranging from 12% to 48%, with N_2 or CO_2 as diluent gas, at two gas temperatures (1130 K and 1650 K). The standoff distance from the coal flame to the burner was used as a metric of ignition delay and the variation in time of the flame location as an indication of the flame stability. It was found that at 1130 K, as oxygen concentration increased, ignition delay decreased. This difference is more evident with N_2 as balance gas than with CO_2. For the same oxygen concentration, the presence of CO_2 was observed to significantly delay ignition at 1130 K. However, a opposite trend, was observed when the gas temperature was 1650 K. In this case, it was observed that a higher oxygen concentration had a detrimental effect on flame stability. This may be due to the high particle temperature at which the char-CO_2 gasification reaction is competing with the char-O_2 reaction. As the gasification reaction is endothermic, this can provoke combustion instability.

3.3 Heterogeneous reactions

Char, the porous solid residue after devolatilisation, consists of extensive, condensed-ring aromatic structures and usually accounts for 30 to 70 wt% of the original coal. It consists of carbon and ash, with small amounts of hydrogen, oxygen, nitrogen, and sulphur. The amount and composition of the char depends on parameters such as parent coal type, pyrolysis temperature, heating rate, pressure, and particle size.

Heterogeneous char oxidation proceeds simultaneously with devolatilisation or after devolatilisation, depending on the reaction conditions. The time required for the combustion of a char particle can be several orders of magnitude greater than devolatilisation. The chemical structure of char does not control the reaction processes to the same extent as devolatilisation, but the physical structure of char, including pore system, surface area, particle size, and inorganic content, plays a significant role. The fundamental surface mechanisms of heterogeneous char gasification and combustion were reviewed by Laurendeau [57], Hurt [58], and Essenhigh [59]. The reactions occur by way of both diffusion and chemical steps as follows:

1. Diffusion of reactant gases (O_2, CO_2, H_2O, H_2) from the bulk gas phase through a relatively stagnant film (boundary layer) to the solid surface of the particle and into the particle's available capillary areas or pore structure
2. Adsorption of reactants on the solid

3. Surface chemical reaction (which may be preceded by reactant dissociation)
4. Desorption of surface reaction products (which contain one or more of the atoms previously part of the solid)
5. Diffusion of the gaseous products into the bulk gas phase

Laurendeau [57] has postulated the existence of three different temperature zones or regimes in which one or more different processes control the overall reaction rate. This "three-zone" theory has been widely accepted and used to interpret experimental data in the char oxidation literature.

According to this "three-zone" theory, char combustion rate is controlled by chemical kinetics at low temperatures (Zone I), oxygen pore diffusion at moderate temperatures (Zone II), and oxygen bulk diffusion at high temperatures (Zone III). However, when all reactants are considered (i.e., O_2, CO_2, H_2O, and H_2), the definition of reaction rate becomes extremely difficult due to the fact that different reactants have different reactivities toward carbon in the char. It is likely that high-temperature gasification with CO_2 or H_2O is kinetically controlled, whereas the O_2 reaction with carbon is in the diffusion-controlled regime at the same temperature. Therefore, the factors influencing the char reactivity and the char reaction rates are discussed in detail in the following sections bellow. Special emphasis is given to char gasification/combustion in CO_2 atmosphere.

3.3.1   Factors influencing the reactivity of coal char

Coal reactivity is affected by different variables that involve the coal properties that cannot be related only to coal physical structure or to process parameters. Coal reactivity decreases as *coal rank* increases. Miura *et al.* [60] found that the gasification reactivity data of chars prepared from 68 different coals by different authors, were very scattered for low-rank coals (carbon content less than 80 wt%), but the reactivity was higher than that obtained for higher-rank coals where the data were less scattered. Several factors concerning the pyrolysis (*thermal history of the char*), such as volatile contents, the temperature at which coal is pyrolysed, the extent of the pyrolysis, the heating rate, and the gas atmosphere at which the pyrolysis occurs, may affect the char reactivity. Char reactivity depends on three characteristics of the given particle, namely, (1) chemical structure, (2) inorganic constituents, and (3) pore structure. Each of these characteristics is discussed briefly here.

Chemical structure

From a molecular point of view, it is necessary to consider the role of active sites when reactivity of carbonaceous materials is studied. Compared with graphitic carbon, the structure of coal char and coke are in more disordered form. The reactivity of disordered carbon is much higher than that of the graphitic carbon because the disordered carbon contains many more edge sites as well as various types of defects that are more reactive than crystalline sites. The complex coal structure requires analysis in terms of its characteristic functional groups.

A typical coal structure consists of aromatic-hydroaromatic clusters (average 2 to 5 rings per cluster) loosely joined together by methylene, ether, and sulphide linkages, 1 to 3 carbon atoms in length [15,57]. The loose aliphatic linkages allow cross-linking between clusters of different planes, and hence the development of an extensive pore structure. Aliphatic, hydroaromatic, and heterocyclic bonds are quite susceptible to bond breakage at pyrolysis temperatures.

Chars are characterised by highly carbon-rich, polynuclear aromatic structures. Edge carbon atoms are at least one order of magnitude more reactive than basal carbon atoms [15]. Increased activity at carbon edges is presumed to be due to the availability of unpaired σ-electrons, which are available to form bonds of chemisorbed species. The σ-electrons of basal carbon atoms are tied up in chemical bonds with adjacent carbon atoms. Furthermore, impurities that may catalyse certain carbon reactions tend to diffuse and concentrate at crystalline edges. Therefore, the difference between "pure" edge and impurity effects becomes hard to distinguish. In the case of carbon reaction with molecular oxygen, several authors [15,61,62] have suggested the existence of oxygen on the basal plane of aromatic structures. This oxygen is considered an additional oxygen source in the $C-O_2$ reaction. Theoretical calculations suggested that when oxygen is placed in the basal plane, the $C-C$ bond strength of the bridging atoms can be reduced by 30%. That means that the reactivity of carbonaceous material will also depend on the capability of trapping oxygen in the basal plane. Oxygen and hydrogen sites also promote char reactivity, since chemisorption on nonaromatic sites is usually favoured over aromatic sites. Oxygen sites are thought to influence reactivity via electron exchanges [57]. Hydrogen sites are presumed to increase char reactivity by preferential oxidation [57].

In general, the chemical structures of the fully devolatilised chars are much more similar than the diverse structure typical for the parent coals [37]. However, the observed reactivity of different coal-rank chars is quite different. These differences in char reactivity can be attributed to variations in the skeletal structure [63], producing variations in active sites, surface area, and pore structure resulting from devolatilisation and particle morphology or to variations in mineral content [64,37]. Consequently, the physical structure and catalytic activity of char minerals are thought to be the main factors controlling the char reactivity.

Inorganic impurities

Inorganic impurities in coal char occur in two forms: mineral matter and trace metals. Mineral matter is typically 10 to 30 wt% in raw coals, consisting of four major types [57]: (1) alumino-silicates (clays) such as kaolinite ($Al_2Si_2O_5(OH)_4$) and illite ($KAl_3Si_3O_{10}(OH)_2$), about 50 wt% of mineral matter; (2) oxides such as silica (SiO_2) and hematite (Fe_2O_3),15 wt%; (3) carbonates such as calcite ($CaCO_3$), siderite ($FeCO_3$) dolomite ($CaMg(CO_3)_2$), and ankerite ($Ca(Mg, Fe, Mn)(CO_3)_2$),10 wt%; and (4) sulphides and sulphates such as pyrite (FeS) and gypsum ($CaSO_4.2H_2O$), 25 wt%.

Typically, mineral matter is randomly distributed in coal as inclusions on the order of 2 μm in diameter. During pyrolysys, gasification, or combustion, mineral matter is transformed to ash (SiO_2, Al_2O_3, FeO_3, CaO, MgO). Some 20 to 30 trace metals are also distributed throughout the coal structure. They can be organically bound to the coal molecule (e.g., boron), bonded inorganically to mineral matter (e.g., zirconium, manganese), or occur in both forms (e.g., copper). Typical concentrations of trace metals in coal are 5 to 500 ppm, although some elements (e.g., B, Ba, Sr, Cu, Mn, Sn, Zr) often appear at the 500-to-1000 ppm level. According to Smooth and Smith [65], the potential effects of mineral matter on char reactivity can be defined as follows: (1) thermal effects, where ash changes the thermal behaviour of char particles and consumes energy; (2) radiation effects, where the radiative properties of ash differ from char and provide a source of radiative heat transfer [66]; (3) particle size effects, where mineral matter affects fragmentation properties; (4) catalytic effects of char reaction; and (5) hindrance effects, where mineral matter may present a barrier through which oxygen must pass to react with carbon and that can impede combustion. Mineral matter and trace elements can provide direct catalytic activity (particularly iron, calcium, and manganese compounds). Laine *et al.* [67] discussed findings of as little as 100 ppm iron that can increase carbon reactivity in CO_2 by factor of 150. There are several theories for the effects of mineral matter and trace metals on char reactivity. The geometric theory suggests that an oxidative intermediate formed by reactant dissociation at a nearby catalytic site migrates to react with carbon. The electronic theory suggests that chemisorption and desorption are favoured (lowering their activation energies) at covalent or ionic carbon-metal bonds generated by electron transfer. However, Mitchell [68] concludes that the catalytic effects due to mineral matter on the burning rate are essential at lower temperatures, but for particle temperatures greater than 1500 K, these effects are negligible.

Pore structure and changes during devolatilisation

The pore characteristic is also a key factor in determining the char gasification/combustion rate, when the reaction is affected by the gas diffusion process. The pore structure of a char particle determines the local concentration of reactant gas molecules within the particle. Pore structure is characterised by three parameters:

- Specific internal surface area, m^2/g
- Specific internal pore volume, m^3/g
- Distribution of internal volume or area over the range of pore diameters, δ

Pore structure is usually classified by considering three broad size ranges: (1) micropores ($\delta \leq 20$ angstroms); (2) mesopores ($20 \leq \delta \prec 500$); or (3) macropores ($\delta \geq 500$). Classification by pore diameter, δ would imply cylindrical geometry of pores. Pore size distribution determines the accessibility of internal surface area to reactant gases. Large surface area of the smallest pores may not be accessible to a reactant unless large feeder pores exist or the reaction kinetics is slow enough to allow time for diffusion into these micropores. Dutta *et al.* [62] found that only pores with $\delta = 20$ to 40 angstroms were available for reaction with CO_2 at 840–1100 °C. Typically, the average pore diameter δ decreases with increasing rank of parent coals. Hippo and Walker [69] studied the reactivity of 16 chars in CO_2 at 900 °C. A char from Pennsylvania low-volatile bituminous coal was over 150 times less reactive than a Montana lignite coal. This finding was attributed to a relative absence of large feeder pores, resulting in poor utilisation of the total surface area. The pore structure of a coal char changes with conversion. Depending on the type of carbon, up to 50% of the volume is initially isolated by micropores and is unavailable for reaction. During early stages of gasification/combustion, specific surface area (and therefore reaction rate) increases. Reactants penetration dictates the development of pore structure as char particles are consumed. Hippo and Walker [69] concluded that the reaction develops new surface area by enlarging micropores but principally by opening up more volume not previously accessible to reactants because the microcapillaries were too small or because existing pores were unconnected. Under slow reaction rates, diffusion of reactants into the micropores favours the development of both micropores and macropores. With higher temperatures, faster reactions will utilise only the most accessible portion of the pore structure, favouring the development of macropores. As the reaction proceeds, surface area and reaction rate increase until the rate of formation of new surface area equals the rate of destruction of old surface area. After that point, surface area tends to remain constant or decrease slightly. The activity of char reaction sites is also a subject to change, either during conversion or as a result of heat pretreatment. Active sites disappear spontaneously due to a thermal healing or annealing of the surface. This annealing process increases rapidly with temperature [70].

Hurt [58] also found that combustion/gasification mainly occurs outside the micropores, that is, on the macropore's surface. He suggested that this is not due to diffusion restrictions, since chars with a larger pore diameter do not have higher reactivity. Instead, he concluded, that there is a higher concentration of active sites in macropores than in micropores. Macropores may appear in crystallite edges or sites in contact with catalytic active inorganic impurities, whereas micropores would be composed of basal planes, which are less reactive. Although the relationship between reaction rate and surface area has been widely studied [70–74], there is no general agreement. Adshiri *et al.* [75] stated that the reaction rate is proportional to surface area. However, most of the studies [71–74] found that surface area and reaction rate are not proportional. Rather, proportionality is found between reaction rate and other parameters such as active surface area (ASA) or coal moisture holding capacity (CMHC). ASA is related to the amount of oxygen chemisorbed by coal, and CMHC is related to the total micropore volume. Parameters like ASA and CMHC are more related to the number of active sites on coal surfaces rather than to the total surface area.

3.3.2 Factors influencing the char reaction rates

In contrast to reactivity, some factors that are solely related to the physical structure of coal or to the conditions in which reactions take place are said to affect the reaction rate.

Reactive gases concentrations

Generally, at low oxygen concentrations, the char oxidation phase of combustion is subsequent to the devolatilisation phase. As the oxygen concentration increases, the ignition temperature falls and ignition heat is reduced. The phases begin to overlap, and with further increase of oxygen concentration the whole combustion process happens progressively faster, with devolatilisation and char oxidation occurring simultaneously [76].

Simultaneous combustion (char-O_2) and gasification (char-CO_2) during air combustion have been studied by Mitchell and Madsen [77] and Wang et al. [78]. The overall burning rates of char particles (diameter: 115 μm) were found to be unaffected by carbon dioxide in the combustion experiments carried out by Mitchell and Madsen [77]. This finding relates to conditions of 3 and 6 vol% oxygen, gas temperature between 1344 and 1507 K, and two concentration levels of CO_2 (2.1 and 8 vol%). Oscillations of particle temperature, with periods ranging from 10 to 100 s, were found in experiments carried out by Kurylko and Essenhigh [79]. These oscillations were considered to be caused by the changing location and intensity of the homogeneous oxidation of CO to CO_2 but, in part, could also have been due to the internal endothermic CO_2 gasification producing CO [79].

Field [80] studied the combustion rate of five size-graded fractions of a low-rank bituminous coal char in the 28–105 μm range at oxygen partial pressures of 0.05–0.1 atm with the particle temperature varying from 1200 K to 2000 K. It was found that the char was attacked by oxygen both externally and internally, the contribution to weight loss from these two being of equal importance. The combustion rate was found to be directly proportional to the oxygen partial pressure (representing first-order reaction, n = 1), and the combustion rate coefficient, based on the external surface area, was independent of oxygen concentration and particle size. It was also found that CO, not CO_2, was the primary combustion product. Later, Field extended his work to 11 different PF chars, swelling and nonswelling, high- and low-rank [81]. The results he obtained showed that the nonswelling coal chars exhibit a uniform decrease in density accompanied by a slight size reduction during conversion, whereas the swelling coal char yielded more scattered results with a marked decrease in density in some cases and a rather steep reduction in size (about the effect of char swelling). The unusual behaviour of the swelling coals was explained as being due to the presence of highly swollen cenospheres in the feed char, which caused the initial density to be very low, and subsequently they burnt faster than the less swollen particles because of their large surface area, thereby causing the density to rise as the median diameter fell due to their complete combustion.

Mulcahy and Smith [82] analysed the internal diffusion and reaction within the porous char and suggested that for small pulverised coal char particles (about 40 μm) of bituminous coal, combustion predominantly occured within the pores, so discussions of the kinetics must be based on penetration (complete or partial) of the particle by the oxidant. They pointed out that for particles larger than 100 μm, the combustion would be mass-transfer controlled in the temperature range of 1200–2300 K. It was emphasised, however, that the rate of combustion becomes increasingly dependent on the chemical reaction rate as particle diameter decreases below 100 μm.

Murphy and Shaddix [83] studied the combustion kinetics of subbituminous coal chars in oxygen-enriched environments by varying the oxygen content from 6 to 36 vol%. As expected, it was found that the oxygen-enriched combustion significantly increased the char combustion temperature and reduced the char burnout time. However, the optical kinetic data, interpreted with a single-film oxidation model, demonstrated an increase of the kinetic control in oxygen-enriched combustion despite the faster particle combustion rates. Fits applied to all the experimental data with both a simple n^{th}-order Arrhenius expression and an nth-order Langmuir-Hinshelwood (L-H) expression yielded apparent reaction orders of 0.2 and 0.3, respectively.

Char gasification is considered a first-order reaction for both CO_2 and steam when working at pressures below atmospheric pressure [71]. For pressures above atmospheric, the reaction order approaches zero. Shufen and Ruizheng [84] found that for lignite coals at 1.6 MPa, reaction orders were 0.26 and 0.34 for steam and CO_2, respectively. Another factor to be considered regarding reactive gases concentration is the inhibition by H_2 and CO. Some studies [85–87] have shown a retarding effect when CO and H_2 are produced. The gasification rate of char-H_2O (R3.6), for instance, decreases almost 40% when the H_2- concentration is the same as that of H_2O.

The CO-inhibition phenomenon has been extensively explained by L-H relations according to the proposed simplified mechanism derived for atmospheric conditions:

$$C_s(\_) + CO_2 \rightleftharpoons C(O) + CO - \Delta H_{1,CO_2} \qquad (3.7)$$

$$C(O) \rightarrow C_s(\_) + CO_x + \Delta H_{2,CO_2} \qquad (3.8)$$

The main characteristic of this mechanism is the inhibition made by the CO produced, which will shift reaction (R3.7) to the left. This effect is negligible for lignite coals and considerable for Pittsburgh bituminous coal chars (up to 10% reduction of the gasification rate) [87]. The influence of CO increases with increase of temperature [88]. Moulijn and Kapteijn [89] considered that the inhibitory mechanism does not fully explain the reduction of gasification rate by H_2. They found experimentally that the gasification reaction stops almost completely when the hydrogen concentration is more than 50 vol%. This suggests that H_2 is part of two different mechanisms during gasification: a reversible reaction that agrees with the L-H kinetics and an irreversible reaction that leads to the deactivation of the active sites.

$$C(O) + H_2O \Leftarrow H_2 + CO + \Delta H_{H_2O} \qquad (3.9)$$

Roberts and Harris [90] have investigated the char conversion rate in a mixture of CO_2 and H_2O. They reported that the reaction rate of a char is not simply the sum of the rates of reaction of char with the pure reactants separately, but the presence of CO_2 reduces (inhibits) the rate of the char-H_2O reaction. The surface areas resulting from reactions in CO_2 are less than those generated under the same conditions in H_2O. With the two reactions occurring in parallel, no increase of the surface area was observed. Hence, the mechanism of inhibition is based on both reactants competing for the same surface reaction sites.

Pressure

Combustion reactivities of pyrolisys and combustion/gasification decrease with increasing pressure [91]. The suppression of tar release under pressure is known to lead to redeposition and repolymerisation in the form of relatively unreactable chars. According to

Messenbroeck *et al.* [50], secondary char deposition appears to affect the combustion reactivities of chars prepared under high-pressure gasification conditions. Monson *et al.* [92] reported that increasing total pressure from 0.1 to 0.5 MPa in an environment of constant gas composition and reactor temperature led to modest increases in the area reactivity, but further pressure increase resulted in decreasing reactivities. Roberts and Harris [93] have measured the intrinsic reactivities of two coal chars to O_2, CO_2, and H_2O at pressures up to 3 MPa and at temperatures whereby chemical reactions alone controlled the conversion rate. They found that the reaction orders in CO_2 and H_2O were not constant over the pressure range investigated, whereas the reaction order in O_2 was unchanged. The activated energies of all three reactions were not found to vary as the pressure was increased. Summarising, there was no significant effect of total pressure on the low-temperature reaction rate of chars with O_2, CO_2, and H_2O. These results indicated that a physical rather than a chemical change is the reason for the observed variations in apparent reaction order. More precisely, it was reported [38,76] that char particles generated at high pressure undergo fragmentation during devolatilisation, thus affecting the structure and the morphology of the char particles and resulting in higher sphericity and a higher surface porosity. It appears that when the reaction is influenced by processes other than surface chemical reaction, the effects of the total pressure on the observed conversion rate can be significant. These effects, however, have little to do with the surface chemical reactions and are more likely to be a result of pressure effects on diffusional processes through the char particles.

3.3.3   Char reaction rates in CO_2/O_2 atmosphere

There are several different ways in which the presence of high concentrations of CO_2 in the bulk gas could influence pulverised char combustion, defined by Shaddix and Molina [94] as follows:

- By hindering the diffusion of oxygen to the char surface, the presence of CO_2 may reduce the burning rate.
- If there is significant heat release in the char particle boundary layer (from oxidation of CO) that is transferring back to the particle, the higher heat capacity of CO_2 may reduce the peak gas temperature and therefore the heat transfer back to the particle, reducing the burning rate.
- Dissociative adsorption of CO_2 on the char surface could result in significant surface coverage and therefore competition for available reaction sites for oxygen, reducing the burning rate.
- Direct gasification of char carbon by CO_2 could contribute to the overall gasification of char surface, increasing the burning rate at a given temperature. However, the endothermicity of the Boudouard reaction would tend to lower the char temperature and thereby lower the overall burning rate.

Shaddix and Molina [94] collected char combustion data of bituminous coals by burning in a gas environment of 12 to 36 mole% O_2 and gas temperatures of 1600–1750 K for both N_2 and CO_2 atmospheres. It was observed that the char particle temperatures were consistently lower in the CO_2 environments, implying a lower overall burning rate. They suggested that the sole influence of CO_2 is through the slower diffusion of O_2 (oxygen diffuses approximately 20% slower in CO_2 than in N_2; see Table 3.1) through the boundary layer surrounding the reacting char particle. Computational analysis showed that the effect of CO_2 on oxygen diffusivity reduces char particle temperatures by approximately 50 K at enriched oxygen levels and decreases the burning rate by approximately 10%. This can be valid, however, for the very high-temperature regimes (Zone III), where the bulk diffusion is the rate-limiting factor. Further, it was demonstrated that the higher molar-specific heat of CO_2 does not significantly influence char combustion. It was also demonstrated that partial CO conversion in the particle boundary layer significantly increases char particle combustion temperatures, especially for oxygen-enriched combustion conditions.

Different results were reported by Wall et al. [95] investigating char reactivity in a drop tube furnace in air and oxyfuel atmosphere as a function of O_2 concentrations. Their results showed that the increase in O_2 concentration leads to increased char burnout for all the coals studied. The results showed also that the char burnout is higher in the oxyfuel atmosphere for almost the entire range (3 to 30 mole%) of oxygen concentrations studied, and this enhancement was attributed to the char-CO_2 gasification reaction. TGA measurements, reported in the same work [95], showed that there was no influence of CO_2 on the char reactivity when the O_2 content in the O_2/CO_2 mixture was varied between 5 and 100 mole%. However, these conclusions may be wrong, since the burnout of the char in these experiments was completed before reaching the gasification commencement temperature of 1073 K. At 2 mole% O_2, the reactivity showed a sharp rise when temperature reached 1073 K. This indicates that there is a competition between O_2 and CO_2 for the gas-char reactions and the char-CO_2 gasification reaction, which could contribute to burnout even at higher O_2 levels if the temperature of combustion is above 1073 K. This behaviour is expected in moderately high-temperature conditions, typical for PF firing, where both diffusion and kinetics are the rate-limiting factors (Zone II).

The effects of coal rank and gas temperature on char-burning rates during oxyfuel combustion were investigated by Shaddix et al. [96]. Five different coals, spanning the rank spectrum from lignite to anthracite, were investigated. The data showed the expected decrease in char reactivity with increasing coal rank. The use of CO_2 diluent yields similar mean particle temperatures as for N_2 diluent at 1130 K. However, at higher gas temperatures (1650 K furnace environment), it was found that for low- and mid-rank coals the char particles burn at lower temperatures when in the presence of CO_2 instead of N_2. The anthracite did not show significant difference in the mean particle temperature measured in N_2 and CO_2. This, according to the authors, can be attributed to the lower diffusivity of oxygen in CO_2 in the particle boundary layer. However, another possible reason can be the lower reactivity of anthracites, thus leading to reduced CO_2-char reaction rates. Schiebahn and Toporov [88] studied numerically char particle conversion rates in oxyfuel atmosphere using kinetic data from different authors. The study considered variation of (1) O_2 content in the O_2/CO_2 bulk, (2) particle temperature, and (3) the char type (10 types from birch char to bituminous). The results showed that the char-CO_2 gasification reaction plays a minor role in low and moderate particle temperatures. At temperatures higher than 1200 K, practically no CO_2 gasification was observed. The most reactive chars (low-ranked coal chars) started to react with CO_2 at about 1500 K with an increasing rate at increasing temperature. A German brown coal char particle (120 μm), for instance, reacts with CO_2 at comparable rates as with O_2 at 1800 K and at about 2 vol% O_2 content in the bulk. Increasing O_2 content reduces the influence of CO_2 on overall conversion rate (at constant particle temperature), but it is still significant, consuming more than 30% of the char carbon with 4 vol% O_2 in the bulk.

Geier et al. [97] have studied experimentally and numerically the influence of oxy-fuel combustion conditions on char combustion rates for several different characteristic pulverised coal particle sizes. According to simulation results, the boundary layer effects became increasingly important for particle sizes greater than 60 μm and reach a maximum for particles around 75 μm in size. For particles larger than 100 μm, the particles burn close to the diffusion limit, and thermal feedback from the boundary layer conversion of CO appears to have a minor impact on the char combustion rate.

3.4   Combustion of volatiles in CO_2/O_2 atmosphere

The substitution of N_2 with CO_2 in the oxidiser will lead to a reduction of the flame speed as reported by Zhu et al. [98]. This causes poor combustion performance and a modified distribution of temperature and species in the combustion chamber. The lower-burning velocity for oxycombustion of gaseous fuel can be theoretically affected by the following features:

- Lower thermal diffusivity of CO_2
- Higher molar heat capacity of CO_2
- Chemical effects of CO_2
- Modified radiative heat transfer

Since the molar heat capacity affects the flame temperature, its effect generally dominates. The lower adiabatic flame temperature in oxycombustion can be increased by increasing the oxygen concentration in the CO_2/O_2 gas mixture, thus reaching similar flame temperature levels as in air combustion [99,100].

Shaddix and Molina [53] investigated experimentally the ignition of coal particles in CO_2 atmosphere. For the conditions of the experiments (large volatile fraction and 100 μm particles), the coal particles undergo homogeneous ignition, with the volatiles being burned in a diffusion flame surrounding the devolatilising coal particle. Thus, the oxygen was prevented from reaching the particle surface during devolatilisation. Before release and ignition of volatiles, the coal particle must be heated by the ambient environment. As discussed in Shaddix and Molina [94], during the initial particle heat-up, the heating rate of a given-size particle is proportional to the product of the gas-particle temperature difference and to the thermal conductivity. Because the thermal conductivity of N_2 and CO_2 at elevated temperatures is practically identical for the experiments conducted, there should be no significant difference in particle heating rates for the two different diluents. Therefore, the initial release of coal volatiles should occur at the same time and at the same rate for the two diluents. Once coal devolatilisation begins, the light hydrocarbons and tar material released during devolatilisation ignite when they come into contact with hot oxidiser gases.

Toporov *et al.* [31] studied experimentally an oil flame (50 kW) with gradual replacement of air as oxidizer with CO_2/O_2 mixture by keeping the O_2 content the same as in air (21 vol%). Increasing the CO_2 concentration in the combusting mixture led to reduction of the reaction rates, shortening the flame, decrease of the flame luminescence, and at 100% replacement to a practical loss of the visible flame. Stabilisation of the flame was obtained only after increasing the O_2 concentration in the combusting mixture, thus demontrating that under oxyfuel conditions, the gaseous flame speed is reduced mostly due to the increased specific heat (thermal diffusivity has a minor role), which reduces the flame temperature and in turn the overall reaction rate.

Andersson and Johnsson [100] studied experimentally a propane flame (80 kW) in air and in CO_2/O_2 mixture. Similarly, the results showed that the temperature levels of the flame with 21 vol% O_2 (oxyfuel-21) are significantly lower than in air due to cooling by the recycled CO_2, which resulted in a suppressed flame development. Compared to the oxyfuel-21 case, an increase of oxygen concentration to 27 vol% O_2 (oxyfuel-27) led to increased temperature levels and improved mixing conditions between fuel and O_2, thus improving the fuel burnout of the oxyfuel-27. The oxyfuel-27 case therefore exhibits similar overall combustion behaviour as the air-fired reference case in terms of gas concentration and temperature profiles.

Liu *et al.* [101] have performed numerical investigations of the chemical effects of CO_2. A comparison between numerical and experimental data showed that the decrease in burning velocity for the oxyfuel combustion cannot entirely be described by considering only the material properties of CO_2. CO_2 affects the combustion reactions especially via the reaction

$$CO + OH \leftrightarrow CO_2 + H, \tag{3.10}$$

which reduces the concentration of important radicals such as H, O, and OH in the combustion chamber and thus decreases the burning velocity. This hypothesis is supported by a comparison of the burning velocity of methane flames and hydrogen flames in a CO_2/O_2 gas mixture. The influence of CO_2 on the burning velocity of hydrogen flames is less significant because the concentration of hydrogen radicals is much higher. Finally, it was summarised, that the chemical effect of CO_2 significantly reduces the burning velocity of methane, whereby the relative importance of this chemical effect increases with increasing CO_2 concentration in the oxidising mixture.

Glarborg and Bentzen [102] did an experimental and theoretical analysis of the formation of CO in oxycombustion of methane and found a strong increase of CO concentrations in the near-burner region. CO_2 competes with O_2 for atomic hydrogen and leads to formation of CO through the reverse reaction of reaction (3.10).

Another consideration for particle ignition when using the CO_2 environment is whether the CO_2 itself can act as a source of radicals via thermal dissociation (into CO + O). Equilibrium calculations, carried out by Shaddix and Molina [53], showed that at the temperature of the experiments (1700 K), dissociation of CO_2 is very minor and therefore is unlikely to influence the volatile ignition delay observed in CO_2 diluent [53]. They also reported that a potential influence of CO_2 on ignition chemistry may stem from reduction of the primary CO oxidation rate in reaction (3.10) and enhancement of key recombination reactions such as $H + O_2 + M = HO_2 + M$ [53]. The chaperon efficiency[1] of CO_2 in these recombination reactions is somewhere between two and three times that of N_2. These chemical kinetic effects are consistent with the observed increase in ignition delay in the CO_2 environments. This topic is addressed in Chapter 5 ,where detailed experimental investigation on methane oxidation in CO_2/O_2 atmosphere is presented.

3.5 Emissions from combustion of pulverised coal in CO_2/O_2 atmosphere

In oxyfuel PF processes, hot gas chemistry and ash composition are different than they are in an air-blown process. Although this aspect has been identified as an R&D requirement for future oxy-fired combustion systems, only a few experimental results have been reported to date. Instead, most data available in the literature are derived from modelling. This may be due to the limited number of research facilities that are able to be operated under the necessary conditions. Furthermore, there are a lot of parameters that have to be investigated, given that oxyfuel firing allows a broad spectrum of variations in, for example, temperature level, oxygen concentration used for firing, and composition of recycled flue gas. An additional degree of freedom arises when constructing the modelled process: where to start recirculation? This, of course, has consequences for the composition of the recycled gas, leading to different subsequent chemical reactions and finally to divergent ash composition and deposits.

3.5.1 NO_x

Two opposing factors affect NO_x emissions during coal combustion: (1) the oxidation of fuel-N by oxygen and other oxidising agents, and (2) the reduction of already produced NO_x by reducing agents, such as hydrocarbons from pyrolysis of the volatile matter in homogeneous reactions and resident char in heterogeneous reactions.

There are numerous technologies available to affect the NO_x emissions produced inside PF-fired furnaces. These methods, known as primary measures and addressed in Chapter 6, range significantly in cost, effectiveness, complexity, and extent of modifications required to achieve the reduction.

Flue gas recirculation (FGR), for instance, uses flue gas recycled from the stack that is injected into the combustion air supply of the burners. Thus, lower oxygen partial pressures in the flame zone are obtained and the peak flame temperature is reduced. The target percentage of recirculated flue gas in the total combustion air flow is typically 10–15%. This method has been used with gas- and oil-fired furnaces. The major effect of flue gas recirculation is on thermal-NO_x formation, with relatively small

---

[1] Enhances third-body reactions.

impact on fuel-NO_x production. Since more than half of the total NO_x produced in a PF-fired furnace will be fuel-NO_x, the most effective measures concentrate on limiting the formation of fuel-NO_x, and therefore the FGR method was rarely considered as an option for NO_x reduction in large coal-fired boilers [103].

The introduction of the oxyfuel technology provides the possibility to reduce NO_x emissions from PF boilers without applying staged combustion but by reconsidering the FGR method. Compared to air-fired units, the NO_x emissions generated per unit of energy during oxyfuel combustion are around 70% lower, depending on the burner design, coal type, and operating conditions [4,104–106]. This substantial NO_x decrease is presumed to be the result of the following mechanisms [107,108]:

- Reduced thermal NO due to the absence of atmospheric N_2
- Reduction of recycled NO after being supplied through the flame: (1) by CH fragments from the pyrolysis of volatile matter; (2) on the char surface, which can be enhanced by an increased CO concentration due to high CO_2 concentrations in the furnace; and (3) by interaction between recycled-NO and released fuel-N, mainly HCN, to form molecular nitrogen.

Liu and Okazaki [109] suggested that possible heat recirculation (e.g., hot recycling) could reduce exhaust NO emission further in three ways: (1) less exhausted flue gas (high recycling ratio); (2) a low oxygen-fuel stoichiometric ratio (SR); and (3) an increase in flame temperature. The concentrations of O_2 in the inlet gas and reducing agents increase simultaneously when decreasing the flue gas recycling ratio, thus affecting the NO_x emissions directly. The conversion of fuel-N to NO_x increases with the O_2 concentration in the mixture, as reported by Hu et al. [108] and Liu et al. [110]. However, an increase in the recycling ratio (lowering O_2 content) leads to an improved NO_x reduction efficiency [107,108]. As the recycling ratio increases, the heterogeneous reaction between NO and char is promoted in the beginning stage by high CO concentrations, which are obtained as a result of the low O_2 content in the inlet gas. However, the resulting decrease of fuel concentration has an inverse effect on the reduction of NO_x because of the decrease in hydrocarbon fragment concentrations as observed by Hu et al. [108]. Hence, there is a contrary effect of O_2 content on the emission of fuel-N and the reduction of recycled-NO_x. The net effect was studied on a real oxycoal flame, and the main results are discussed in Chapter 7.

Park et al. [111] studied the fuel-N conversion during heterogeneous reaction of bituminous coal char with O_2, CO_2, and H_2O over a broad range of temperatures, pressures, and reactant gas concentrations. The results showed that char-N is converted as follows: entirely to N_2 when char reacts with CO_2; to N_2 and NO when char reacts with O_2; and to HCN, NH_3, and N_2 when char reacts with H_2O.

Shaddix and Molina [112] performed experiments in a down-fired entrained flow reactor measuring NO_x formation at three different oxygen concentrations (12%, 24%, and 36%) in both N_2 and CO_2. For enhanced oxygen levels, combustion in a CO_2 environment led to a decrease of the NO_x formation. This was explained with the lack of thermal-NO_x production in the N_2-free gas environment and with lower volatile flame temperatures and lower char combustion temperatures in a CO_2 environment.

When bituminous coal particles are burned in an atmosphere of 12% O_2, only 1/5 of the fuel-N_2 is converted to NO_x. At 24% O_2, approximately 1/3 of the fuel-N_2 is converted to NO_x in a N_2 diluent and about 1/4 is converted when in a CO_2 environment. At 36% O_2, the conversion factors increase to 55% and 45%, respectively. For highly volatile coals, the fuel nitrogen conversion factor was approximately 35% for combustion in 12% O_2, 60% for combustion in 24% O_2, and 80% for combustion in 36% O_2, with similar decreases in conversion factor in CO_2 environments as obtained for the bituminous coal.

The results showed also that volatile-generated NO_x is a larger fraction of the total coal-NO_x generated at enhanced oxygen levels. This may result from the enhanced effect of oxygen on the volatile flame temperature compared to char combustion temperature. The influence of CO_2 on char-NO_x formation appears to be of a similar relative magnitude at 24% O_2 and 36% O_2 but appears to have almost negligible influence at 12% O_2, as was also found during combustion of the raw coal particles.

When NO_x is recycled back to the flame at 12% O_2, net NO_x destruction during coal combustion was observed, and at 24% O_2 there is essentially zero net NO_x production. However, at 36% O_2, there is significant net production of NO_x for both N_2 and CO_2 diluents [112].

3.5.2 SO_x

Compared to operation with air, SO_2 emissions are also reduced in oxy-firing, and this is related to the increased capture of SO_2 by the ash as a result of its altered chemistry. Croiset and Thambimuthu [4] observed that the conversion of fuel-S into SO_2 is independent of the oxygen concentration. However, the type of environment (air, O_2/CO_2, and O_2/RFG) has some impact on the conversion. The conversion drops from 91% for air operation to an average of 75% during oxy-firing in a CO_2/O_2 mixture and to 64% for recycled combustion. A possible explanation for this is that a fraction of SO_2 is further oxidised to SO_3, thus leading to high SO_3 concentrations in the flue gas. Comparisons between dry and wet recycle tests show no significant difference in SO_2 emission rate, which indicates that the removal of SO_2 through water condensation (dry recycling) is not significant. However, the SO_2 concentrations are higher than the ones seen with air combustion due to the gas recycling, with the measured levels depending on the recycle ratio (O_2 concentration in the oxidiser mixture).

3.5.3   Ash composition

With regard to ash composition, e.g., Wigley and Goh [113] reported on the collection of a large portfolio of ash deposits from oxy-firing trials with a 1 MW_{th} combustion test facility. Combustion parameters that have been varied include firing mode (air or oxyfuel), excess oxygen levels, the proportion of air staging, and the degree of oxygen enrichment. As a result of these variations, very different fly ashes were obtained. This is probably due to the chosen firing conditions, which led to different temperatures during combustion. Gas and particle composition affects corrosion and particle removal. As pointed out by Schnurrer et al. [114], higher SO_3 concentration will assist electrostatic precipitators (ESPs); by contrast, the formation of alkali sulphates will enhance corrosion. Therefore, alkali reduction was investigated under oxyfuel conditions, for example, by Müller [115]. The lack of experimental data becomes very evident in this area of study, since most of the reported results were obtained by computations that assume chemical compounds to be at equilibrium. Due to the high partial pressure of CO_2 in these systems, however, the reaction kinetics may play a dominant role for the resulting chemistry. For example, in investigating alkali-sulphate/-carbonate deposits during oxy-firing, Kellermann et al. [116–117] observed that the carbonate species dominate by far in fly-ash deposits.

From the current state of the research, the urgent necessity for experimental data becomes evident. But in contrast to air-blown systems, the unique possibility arises to vary a whole set of parameters, e.g., temperature level, the oxygen concentration used for firing, and the composition of recycled flue gas.

3.6 Heat transfer in RFG/O_2 atmosphere

By recycling CO_2 (and possibly H_2O and SO_2, if the recycled gas is not dried) from the flue gas outlet back to the furnace inlet, the heat transfer inside the furnace will be modified due to:

- The convective heat exchange that will be affected by the changed gas heat capacity and gas temperature
- The radiative heat flux that will be affected by the changed gas emissivity and/or absorptivity

The major contributor of heat transfer from a pulverised fuel flame in conventional air combustion is the thermal radiation from hot particles (fly ash, soot, and char) and from products of combustion such as water vapour, carbon dioxide, sulphur dioxide, and carbon monoxide [66,118]. To understand the exact impact of the increased partial pressures of CO_2 and water vapour in O_2/RFG PF combustion on the heat transfer, special attention to the radiative properties of the gas is given below.

3.6.1 Radiative properties of CO_2 and water vapour

Although particles emit in the whole spectrum, the gases emit and absorb the radiant energy only in narrow wavelength or frequency bands. Therefore, it can be expected that at certain wavelengths these gases will be completely transparent. Thus an engineering model can be developed by concentrating only on the part of the spectrum in which the combustion gases participate. The width of each of these so-called wide bands is an adjustable parameter and it is determined by assuming that the absorption and emission of radiation in the band are equal to the effective absorption and emission of the spectral lines presented within the band. From theory it is known that the size of the absorption band depends on temperature, partial pressure of absorbing gases, and path length, as shown in Figures 3.1 and 3.2 for CO_2 and water vapour, respectively.

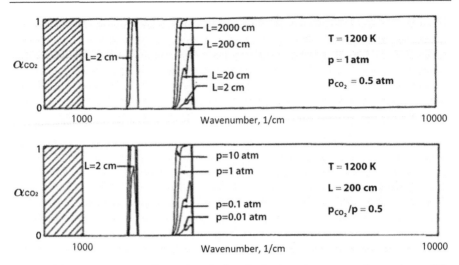

Figure 3.1 CO_2 spectra: Effect of path length (top) and pressure (bottom) on CO_2 absorptivity [119].

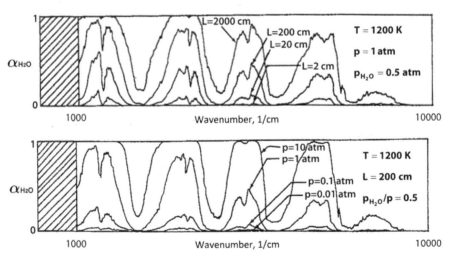

Figure 3.2 Water vapour spectra: Effect of path length (top) and pressure (bottom) on H_2O absorptivity [119].

This dependence can be summarised as:

- The effect of an increase of the path length and the partial pressure is similar and both lead to an increase of the spectral absorptivity α_{H_2O} and α_{CO_2}. This is valid not only inside the boundaries of their absorption widebands but also to a broadening of these bands in both directions of the wavelength spectrum, with a much stronger effect contributed by H_2O, which, for boiler typical dimensions (optical path lengths), leads to an almost continuous spectrum of emission.

• Changes of the partial pressure and path lengths have a different effect on α_{H_2O} and α_{CO_2}. In the case of CO_2, these parameters have a similar influence, and therefore α_{CO_2} can be given as a function of CO_2 partial pressure and path length $p_{CO_2}L$. However, in the case of water vapour, the partial pressure has a stronger effect on α_{H_2O} than a change of the path length. Therefore, even at the same value of $p_{H_2O}L$ but different p_{H_2O} and L, the spectral absorptivity is different. The change of the partial pressure leads to changes in the absorption coefficient. Because the mean distance between the spectral lines for CO_2 is much smaller than for H_2O, this change is less significant for CO_2.

The total emissivities of the two main gases CO_2 and H_2O, obtained by different authors, were summarised by Blokh [66]. He derived that for a given pL, water vapour has higher emissivities compared to CO_2. The general trend is that the emissivities of both gases decrease with an increase in temperature above 1000 K and increase with an increase of pL - partial pressures of the gases and the optical length. The temperature dependence of emissivity, however, is somewhat different for CO_2 and water vapour. For CO_2 it follows a behaviour of $\epsilon_{CO_2} \approx 1/\sqrt{T}$ and for water vapour $\epsilon_{H_2O} \approx 1/T$. Overall, the emission of CO_2 relates to $T^{3.5}$ and for water vapour to $T^{3.0}$.

During oxy-firing the flue gas consists mainly of CO_2 and water vapour. Therefore, taking into account the role of these triatomic gases in the radiative heat transfer, the impact of the gaseous matter on the flame characteristics has to be considered.

3.6.2   Dry and wet recycle and the effect on the radiative heat transfer

The effect of propane oxy-firing with 27 vol% O_2 in a CO_2/O_2 mixture on flame emissivity was studied by Andersson and Johnsson [100], who showed that it led to an increase of the flame emissivity of up to 30% compared to an air-propane flame. This is remarkable, since the temperature levels were generally lower or similar to those of the air case. However, the measured difference in total mean emissivity compared to the difference in gas emissivity for the air and oxy-firing case shows that the obtained higher-radiation intensity in the oxy-firing case can be explained not only by an increase of the gas emissivity but also by an increase of the soot volume fraction and the related emission from soot.

For PF oxy-firing (lignite) in a CO_2/O_2 atmosphere, Andersson et al. [3] observed that the gas would contribute with 35–40% to the radiation impact on the combustor wall in the investigated lab-scale 100 kW$_{th}$ oxyfuel case, whereas it would be less than 30% in the air-fired case under similar temperature conditions. However, the increase in gas radiation incident on the furnace walls is only moderate due to the absorption by the small gas layer between the high-temperature flame region and the combustor walls. Hence, the total intensities measured in this bench-scale furnace were similar for air-PF and oxyfuel conditions as long as the temperatures were similar. Based on these observations, Johansson [120] introduced a WSGG-modified model based on a "four-gray-one-clear" model for describing the absorption coefficient in oxy-firing. The advantage of this model is that it considers the changed gas properties in oxy-firing conditions while using the fast computational approach provided by the original

WSGGM. Similar approaches were adopted by Gupta *et al.* [121] and Payne *et al.* [122], who showed, by calculating the radiative heat transfer from a flame resulting from oxy-firing, that the original "three-grey-one-clear gas" model should be validated and/or modified or replaced by a more accurate model. A new "four-grey-one-clear" gas model was introduced by Gupta *et al.* [121] and validated against Edwards's EWBM in simulations of large-scale furnaces. Payne *et al.* [122] introduced a "seven-grey-gas" model, which was validated against measurements in a 117 kWth bench-scale combustor and in a 3 MWth pilot-scale oxycoal firing furnace. In this case, the emissivity increased from $\epsilon = 0.16$ for air to $\epsilon = 0.25$ for operation with an oxidiser consisting of a dry CO_2/O_2 mixture that contained 31 vol% O_2. An extrapolation made with this model showed that the gas mixture emissivity at 1600 K increases in the pilot scale and in the boiler furnace from $\epsilon = 0.43$ to $\epsilon = 0.54$ for a similar change in operation. Payne *et al.* [122] also investigated the optimum conditions for a 50 MWe retrofit furnace, assuming the same heat transfer as for air operation in the furnace and at the convective part of the furnace. They reported that the predicted optimum recycle ratios required to achieve a heat transfer efficiency that is comparable to operation in air correspond to an O_2 concentration in the O_2/RFG mixture of approximately 23.8 vol% for wet recycle and 27 vol% for dry recycle conditions. Finally, it was concluded that because the emissivity of the boiler furnace is larger than in bench- and pilot-scale systems, a stronger decrease in adiabatic flame temperature is required to compensate for the gain in efficiency. Consequently, a lower O_2 concentration in the oxidising mixture is required. Hence, in an industrial unit size, which normally has a high surface firing density due to the increased particle radiation, the furnace volume emissivity gradually approaches one and performance is no longer sensitive to changes of flue gas composition.

Erfurth *et al.* [123] simulated the combustion of bituminous coal in a 1200 MWth oxy-firing furnace. For this study, a nongrey implementation of the EWBM model was used, as proposed in Erfurth *et al.* [124]. The predictions showed that choosing the recycling ratio in oxy-firing based on a constant adiabatic flame temperature (27 and 30 vol%. O_2 for wet and dry recycling, respectively) leads to flame temperatures comparable to air combustion. Similar temperature fields, however, are associated with an increase of the incident surface radiation of 14% and 6% for wet and dry recycling, respectively. Accordingly, the authors suggested that a constant heat flux to the furnace walls can be achieved at oxygen concentrations in the O_2/RFG mixture equal to 23.8 vol% for wet and 28.6 vol% for dry recycling.

Such predictions should be taken into account before it becomes possible to perform large-scale boiler measurements in order to estimate the exact effect of the flue gas on the incident radiation to the furnace walls.

3.6.3   Dry and wet recycle and the effect on the convective heat transfer

The increase in thermal capacity of the flue gas in oxy-firing increases the heat transfer in the convective section of the boiler compared to air firing. When oxy-firing with an oxygen concentration higher than that in air (21 vol%) is realised, the volume of gas passing through the boiler in the oxyfuel case is lower. Therefore, the heat transfer in the radiative and convective sections of the boiler will need to be optimised to ensure efficient operation. Payne *et al.* [122] suggested the following simple model for the convective heat transfer coefficient:

$$h_1 = h_0 \left(\frac{Re_1}{Re_0}\right)^m \left(\frac{Pr_1}{Pr_0}\right)^n \left(\frac{k_1}{k_0}\right) \tag{3.11}$$

where k denotes the thermal conductivity, index 0 refers to the PF-air system, and index 1 refers to the PF-O_2/RFG system. For the walls of the furnace a value of $m = 0.8$ was assumed; for the heat exchangers in the upper furnace and the convective passages a value of $m = 0.6$ and in both cases $n = 0.33$ was used. In case of a retrofit, where global furnace heat transfer is matched and a given flue gas oxygen concentration of 3%, which is typical for air firing, can be assumed, the oxy-firing case will lead to moderate changes in relative heat flux distribution to the various heat exchanger sections of the boiler that must be considered. These changes are more pronounced for wet recycle than for dry recycle. This is due to the higher water vapour content in flue gases, which leads to an increase in convective heat transfer in the low-temperature heat exchangers ($Pr \sim c_p$). Hence, a considerable amount of attemperation for both superheated and reheated steam was suggested by Payne *et al.* [122].

3.7 Summary of Chapter 3

A detailed literature study has been carried out, reviewing the basic mechanisms of particle combustion with special attention given to combustion in an atmosphere with a high content of CO_2. The following issues can be identified:

1. Relating to particle devolatilisation and ignition:

   - The presence of CO_2 retards devolatilisation and single coal particle ignition.
   - The flame propagation velocity of a PF cloud in CO_2/O_2 mixture is just 20 to 30% of that in air at the same O_2 concentrations.
   - Ignition in CO_2/O_2 mixture is homogeneous.
   - An increase of O_2 partial pressure in the CO_2/O_2 mixture can lead to similar reaction rates as in air. However, at higher temperatures, this can have an opposite effect.
   - Simultaneous occurrence of the thermal cracking and CO_2 gasification of the nascent char, whereas the rates of the gasification seem to be much higher than those reported previously.

2. Relating to heterogeneous reactions:

   - At least three heterogeneous reactions take place simultaneously, namely char-O_2, char-CO_2, and char-H_2O.

- The reaction char-O_2 is dominant, but the contribution of the gasification reactions becomes significant at high (flame) temperatures and at low O_2 concentrations.
- Since gasification reactions are slower than the char-O_2 reaction, they are mainly kinetically controlled at typical combustion temperatures.
- The rate of reaction of a char is not simply the sum of the rates of reaction of char with the pure reactants separately but rather the presence of CO_2 reduced (inhibits) the rate of the char-H_2O reaction by competition of both reactants for the same active surface sides.
- The presence of CO and H_2 in the bulk leads to inhibition effects on char-CO_2, and char-H_2O reactions, respectively.
- The gasification rates (char-CO_2 and char-H_2O reactions) depend not only on CO_2 and H_2O partial pressures but also on the coal quality.

3. Relating to homogeneous reactions:

- The presence of CO_2 retards the chemical reactions due to its higher heat capacity.
- The presence of CO_2 retards the CO oxidation due to chemical effects.

4. Relating to emissions:

- Reduced thermal NO due to absence of atmospheric nitrogen.
- Reduction of recycled NO after being fed back into the flame: (1) by CH fragments from the pyrolysis of volatile matter; (2) on the char surface, which can be enhanced by an increased CO concentration due to high CO_2 concentrations in the furnace; and (3) by interaction between recycled-NO and released fuel-N, mainly HCN, to form molecular nitrogen.
- SO_2 concentrations are higher than in air-firing in case of wet recycle.

5. Relating to heat transfer:
- The increase of partial pressures of CO_2 and steam in flue gas increases the flue gas absorptivity/emissivity.
- The effect of steam on radiative heat transfer is more pronounced than that of CO_2.

Thus, based on the first experience accumulated worldwide, the flame stability and the related design of a PF burner, the formation of pollutants and the measures for their control, and the radiative heat transfer were identified as the main challenges in the process of PF combustion in a mixture of oxygen and recycled flue gas. These questions are addressed in Part II.

4 Mathematical Modelling and Model Validations

Computational methods in heat transfer and fluid flow have been advanced to the point that such methods currently are widely applied to the simulation of complex—yet practical—combustion systems. The designer uses these computational tools to accurately determine performance at the conceptual design stage.

The simulation of pulverised fuel (PF) combustion systems is based on a complex mathematical model, which includes modelling of the fluid flow, turbulence, chemical reactions and radiative heat transfer in an Eulerian framework, and modelling of the coal particles transport, heterogeneous reactions, and the associated momentum, heat, and mass transfer with the surrounding reacting fluid in a Lagrangian framework. Describing these phenomena is quite a complex matter, and intensive research in this area is still going on.

The first computational method for PF combustors was published by Gibson and Morgan [125]. Later, a number of more sophisticated models were introduced in pioneer works, such as those from Smoot and Pratt [126], Lockwood *et al.* [127], Truelove [128], Fiveland *et al.* [129], Williams *et al.* [130], Niksa [131], Weber and co-workers from IFRF [132], the review book edited by Smoot [15], the Ph.D. theses of Schnell [133] and Azevedo [134], and so on.

This chapter presents a numerical model for PF oxyflames. The discussions here are particularly directed to turbulence combustion (Section 4.1), coal particle combustion (Section 4.2), and gas emissivity (Section 4.3) submodels used in simulations of PF combustion processes with respect to the necessary modifications required for oxycoal simulation applications. In addition, the modified coal particle models are validated against experimental data obtained for a variety of boundary conditions.

4.1 Turbulent combustion modelling

The Eulerian part of the model for a two-phase reacting turbulent flow can be expressed as a system of partial differential equations accounting for the conservation of mass, momentum, energy, and chemical species. The following system presents the governing equations in Favre-averaged[1] quantities as follows:

---

[1]Density-weighted Favre averaging is normally introduced for systems with large changes in density, such as combustion. The dependent variable Φ can be decomposed into a mean part $\widetilde{\Phi}$ and a fluctuating part Φ'' as follows: $\Phi = \widetilde{\Phi} + \Phi''$.

Combustion of Pulverised Coal in a Mixture of Oxygen and Recycled Flue Gas. http://dx.doi.org/10.1016/B978-0-08-099998-2.00004-7

The continuity equation:

$$\frac{\partial \bar{\rho}}{\partial t} + \frac{\partial}{\partial x_j}(\bar{\rho}\tilde{u}_j) = S_m \tag{4.1}$$

The momentum transport equation:

$$\frac{\partial(\bar{\rho}\tilde{u}_j)}{\partial t} + \frac{\partial}{\partial x_j}(\bar{\rho}\tilde{u}_i\tilde{u}_j) = \overline{\rho f_1} - \frac{\partial \bar{p}}{\partial x_i} + \frac{\partial}{\partial x_j}\left(\bar{\tau}_{ij} - \overline{\rho u_i'' u_j''}\right) + S_{u_j} \tag{4.2}$$

In most combustion systems, significant density variations arise due to the heat release. Therefore, the conservation equations for energy and for the mass of individual species must be solved in order to obtain the density field.

The enthalpy transport equation:

$$\frac{\partial(\bar{\rho}\tilde{h})}{\partial t} + \frac{\partial}{\partial x_j}(\bar{\rho}\tilde{u}_j\tilde{h}) = \frac{\partial}{\partial x_j}\left(\Gamma_h \frac{\partial \tilde{h}}{\partial x_j} + \Gamma_h \frac{\partial \tilde{h}''}{\partial x_j} - \overline{\rho u_j'' h''}\right) + \bar{\rho}\tilde{S}_{ch} + S_h + S_r \tag{4.3}$$

The chemical species mass fractions transport equation:

$$\frac{\partial(\bar{\rho}\tilde{Y}_k)}{\partial t} + \frac{\partial}{\partial x_j}(\bar{\rho}\tilde{u}_j\tilde{Y}_k) = \frac{\partial}{\partial x_j}\left(\Gamma_Y \frac{\partial \tilde{Y}_k}{\partial x_j} + \Gamma_Y \frac{\partial \tilde{Y}_k''}{\partial x_j} - \overline{\rho u_i'' Y_k''}\right) + \bar{\rho}\tilde{\omega}_k + S_{Y_k} \tag{4.4}$$

with N different reacting species; $N - 1$ equations are needed. The last mass fraction (usually the bigger one) can be calculated from the relation $\sum_k Y_k = 1$.

The equation of state is:

$$\bar{\rho} = \frac{\bar{p} + P}{R\tilde{T} \sum_k \frac{\tilde{Y}_k}{M_k}} \tag{4.5}$$

where R is the universal gas constant and M is the mean molecular weight of the mixture.

Here S_m, S_{u_j}, S_h, S_{Y_k} are the extra source terms of mass, momentum, energy, and chemical species k due to exchange with the discrete phase. The source terms S_r and S_{ch} appearing in the enthalpy equation represent the heat source due to radiation and to chemical reactions, respectively.

The sensible enthalpy of mixture is defined by the expressions:

$$\tilde{h} = \sum_{k=1}^{K} \tilde{Y}_k \tilde{h}_k \tag{4.6}$$

where $\tilde{h}_k = \int_{T_{ref}}^{T} c_{p,k} dT$, with T_{ref} set at 298 K.

The specific heat capacity, c_p, of a gas mixture at constant pressure is defined as:

$$c_p = \sum_{k=1}^{K} \widetilde{Y}_k c_{p,k} \tag{4.7}$$

with the specific heat capacity, $c_{p,k}$, of species k at constant pressure being calculated using a temperature-dependent polynomial for each species k.

The above system includes unknown terms that can be divided into two groups:

1. Terms defining the diffusive fluxes $\overline{\rho u_i'' u_j''}$, $\overline{\rho u_i'' Y_k''}$, $\overline{\rho u_i'' h''}$

2. Terms defining the mean reaction rate $\overline{\rho} \widetilde{\omega}_k$

How these terms influence mean values is a key issue in turbulent combustion modelling. In order to close the system of equations (Eqs. 4.1–4.5), these terms require closure assumptions and models to express them as functions of the solved mean values of the basic variables ($\overline{\rho}, \widetilde{u}_i, \widetilde{h}$ and $\widetilde{Y}_k$) or as solutions of additional conservation equations.

Models used in the current study for the diffusive fluxes and for the mean reaction rate are discussed in Sections 4.1.1 and 4.1.2, respectively.

4.1.1 Diffusive fluxes

The terms of group 1 above represent the transport of momentum, species, or energy due to turbulent fluctuations. The above system of equations cannot be solved directly, and therefore the turbulent correlations $\overline{\rho u_i'' u_j''}$, $\overline{\rho u_i'' Y_k''}$, $\overline{\rho u_i'' h''}$ need to be determined, e.g., modelled.

The simplest method is to assume gradient transport, which for the Reynolds stresses will take the form:

$$\overline{\rho u_i'' u_j''} = \mu_t \left(\frac{\partial \widetilde{u}_i}{\partial x_j} + \frac{\partial \widetilde{u}_j}{\partial x_i} \right) - \left(\frac{2}{3} \overline{\rho} \widetilde{k} + \frac{2}{3} \mu_t \frac{\partial \widetilde{u}_k}{\partial x_k} \right) \delta_{ij}, \tag{4.8}$$

where μ_t, the dynamic turbulent viscosity, is to be obtained through modelling in some fashion (based on the Boussinesq's theory). In this work, μ_t is found from

$$\mu_t = C_\mu \overline{\rho} \frac{\widetilde{k}^2}{\widetilde{\epsilon}}, \tag{4.9}$$

where C_μ is a model constant. From Eqs. (4.8) and (4.9) it is clear that the Reynolds stresses $\overline{\rho u_i'' u_j''}$ can be found if the mean turbulent kinetic energy $\widetilde{k}$ and its mean dissipation rate $\widetilde{\epsilon}$ are known. This is the basis of the k-ϵ turbulence model, where transport equations for $\widetilde{k}$ and $\widetilde{\epsilon}$ are solved. In the present work, these equations were modelled as follows [135]:

The model transport equation for the turbulent kinetic energy:

$$\frac{\partial (\overline{\rho} \widetilde{k})}{\partial t} + \frac{\partial}{\partial x_j} \left(\overline{\rho} \widetilde{u}_j \widetilde{k} \right) = \frac{\partial}{\partial x_j} \left[\left(\mu + \frac{\mu_t}{\sigma_k} \right) \frac{\partial \widetilde{k}}{\partial x_j} \right] + \overline{\rho} \left(P_k - \widetilde{\epsilon} \right) + \Pi_k, \tag{4.10}$$

where σ_k is a model constant and the rate of turbulent kinetic energy production P_k is closed in the same way as the Reynolds stresses are already modelled as follows:

$$P_k = \frac{\mu_t}{\bar{\rho}} \left(\frac{\partial \tilde{u}_i}{\partial x_j} + \frac{\partial \tilde{u}_j}{\partial x_i} \right) \frac{\partial \tilde{u}_j}{\partial x_i} - \left(\frac{2}{3} \bar{\rho} \tilde{k} + \frac{2}{3} \mu_t \frac{\partial \tilde{u}_k}{\partial x_k} \right) \delta_{ij}. \tag{4.11}$$

The pressure term, Π_k, is very often neglected, since a generic and fully accepted model is still to be developed. The current study does it also. The model transport equation for the dissipation rate of turbulent kinetic energy can be written as:

$$\frac{\partial (\bar{\rho} \tilde{\epsilon})}{\partial t} + \frac{\partial}{\partial x_j} \left(\bar{\rho} \tilde{u}_j \tilde{\epsilon} \right) = \frac{\partial}{\partial x_j} \left[\left(\mu + \frac{\mu_t}{\sigma_\epsilon} \right) \frac{\partial \tilde{\epsilon}}{\partial x_j} \right] + \bar{\rho} \frac{\tilde{\epsilon}}{\tilde{k}} \left(C_{\epsilon 1} \tilde{k} P_k - C_{\epsilon 2} \tilde{\epsilon} \right) \tag{4.12}$$

where σ_k, $C_{\epsilon 1}$ and $C_{\epsilon 2}$ are model constants. The terms, as shown here, follow the "standard" k-ϵ turbulence model for high-Reynolds-number flows proposed by Launder and Spalding [135]. Numerical values for the various model constants are given in Table 4.1. The model of Launder and Spalding (with small or no changes) and closely related models are widely applied in research and industry today. The application of the k-ϵ model, however, requires very careful attention concerning its abilities to model turbulent flows. The relation between strain rates and stresses, for instance, given in Eq. (4.8), is the same, regardless of direction. In addition, only the strain rates have direction, whereas the turbulence viscosity is the same in all directions. Hence, some effects, such as strong curvature in the flow (e.g., recirculation zones, swirl) or strong anisotropy in the Reynolds stresses as well as directional forces that influence the turbulence (e.g., buoyancy), may cause difficulties with k-ϵ and other two-equation models employing the Boussinesq's eddy viscosity concept.

An alternative to the gradient transport hypothesis, Eq. (4.8), is to solve a transport equation for each of the Reynolds stresses. However, in the exact transport equations for Reynolds stresses, unknown correlations of the third order appear, and therefore modelling assumptions are needed to close the system of equations. The resulting Reynolds stress equation has the form:

$$\frac{\partial (\overline{\rho u_i'' u_j''})}{\partial t} + \frac{\partial}{\partial x_k} \left(\overline{\rho u_i'' u_j''} \tilde{u}_k \right) - \bar{\rho} \left(D_{ij,v} + D_{ij,t} \right) = \bar{\rho} \left(P_{ij} + \Phi_{ij} + \epsilon_{ij} \right), \tag{4.13}$$

where the first two terms represent the transient and the convective transport with the mean flow; $D_{ij,v}$ and $D_{ij,t}$ are the viscous diffusion and turbulent diffusion; P_{ij} is production (work conducted by the Reynolds stresses $\overline{u_i'' u_j''}$, transfer of mechanical energy from the mean flow to the turbulence); Φ_{ij} is redistribution (exchange of

Table 4.1 The values of the constants in the k-ϵ turbulence model.

C_μ	$C_{\epsilon 1}$	$C_{\epsilon 2}$	σ_k	σ_ϵ
0.09	1.44	1.92	1.0	1.3

energy between the components); and ϵ_{ij} is dissipation (transfer of kinetic energy to thermal energy). The convection, the production P_{ij}, and the viscous diffusion $D_{ij,v}$ are functions of known quantities and do not require modelling:

$$\rho D_{ij,v} = \frac{\partial}{\partial x_k}\left(\mu\frac{\partial\overline{u_i''u_j''}}{\partial x_k}\right) \tag{4.14}$$

$$P_{ij} = -\overline{u_k''u_j''}\frac{\partial\tilde{u}_j}{\partial x_k} - \overline{u_k''u_i''}\frac{\partial\tilde{u}_j}{\partial x_k} \tag{4.15}$$

The rate of turbulence energy production P_k used in Eq. 4.12 is now given by:

$$P_k = \frac{1}{2}P_{ii} \tag{4.16}$$

In addition, a length scale or timescale for the turbulence has to be defined. The usual way to provide such a scale, is to solve an equation for the dissipation rate of turbulent kinetic energy, $\tilde{\epsilon}$. This can be the same $\tilde{\epsilon}$-equation as Eq. (4.12) or with a modified turbulence diffusion term:

$$D_{\epsilon,t} = \frac{\partial}{\partial x_j}\left(C_\epsilon\frac{\tilde{k}}{\tilde{\epsilon}}\overline{u_j''u_k''}\frac{\partial\tilde{\epsilon}}{\partial x_k}\right) \tag{4.17}$$

The model constant C_ϵ is often set to 0.18. Here, $\tilde{k}/\tilde{\epsilon}$ is used as timescale.[2] The turbulence kinetic energy can be found from the impression:

$$k = \frac{1}{2}\overline{u_j''u_k''} = \frac{1}{2}\left(\overline{u_1''^2} + \overline{u_2''^2} + \overline{u_3''^2}\right) \tag{4.18}$$

The remaining terms Φ_{ij}, $D_{ij,t}$, and ϵ_{ij} in Eq. (4.13), however, need to be modelled in order to close the equation. The term Φ_{ij} represents the pressure-strain correlations and is often called redistribution.

$$\rho\Phi_{ij} = \overline{p'\left(\frac{\partial u_i''}{\partial x_j} + \frac{\partial u_j''}{\partial x_i}\right)} \tag{4.19}$$

The classical approach to modelling the redistribution term is to divide it into three components:

$$\Phi_{ij} = \Phi_{ij,1} + \Phi_{ij,2} + \Phi_{ij,w}, \tag{4.20}$$

where the so called "slow" component is modelled as:

$$\Phi_{ij,1} = -C_1\frac{\tilde{\epsilon}}{\tilde{k}}\left(\overline{u_i''u_j''} - \frac{1}{3}\delta_{ij}\overline{u_k''u_k''}\right) \tag{4.21}$$

[2]It expresses the "lifetime" of a turbulence eddy.

and the "rapid" component is modelled as:

$$\Phi_{ij,2} = -C_2 \left(P_{ij} - \frac{1}{3}\delta_{ij} P_{kk} \right) \tag{4.22}$$

The model for the rapid term $\Phi_{ij,2}$ was originally proposed as an alternative of the model for the slow term, $\Phi_{ij,1}$. It has appeared, however, that the sum of these two models gives reasonably good results [136].

Two sets of constants have gained status as a kind of standard: Gibson and Lauder [137]: $C_1 = 1.8$ and $C_2 = 0.6$; and Gibson and Younis [138]: $C_1 = 3.0$ and $C_2 = 0.3$.

The third term, $\Phi_{ij,w}$ is caused by reflection of pressure fluctuations from the wall when wall-flow interactions are considered and models the damping of the fluctuations normal to the wall. This term has an effect far into the flow from the wall as well as outside the boundary layer [136].

The redistribution model is regarded as the most uncertain part of the model equation for the Reynolds stresses. Although new and better models are under development [139], it seems that the redistribution model is "far ahead to a final model," in particular with combustion in mind [136].

The most common model for turbulence diffusion term $D_{ij,t}$ is a gradient model according to Daly and Harlow [140]:

$$D_{ij,t} = \frac{\partial}{\partial x_k}\left(C_s \frac{\tilde{k}}{\tilde{\epsilon}} \overline{u_k'' u_l''} \frac{\partial \overline{u_i'' u_j''}}{\partial x_k} \right) \tag{4.23}$$

The model constant C_s is often set to 0.22. Although the model of Daly and Harlow is based on gradient diffusion only, it is considered a standard or basic model for turbulent diffusion. It can be further simplified by use of a turbulent viscosity:

$$D_{ij,t} = \frac{\partial}{\partial x_k}\left(C_\mu' \frac{\tilde{k}^2}{\tilde{\epsilon}} \frac{\partial \overline{u_i'' u_j''}}{\partial x_k} \right) \tag{4.24}$$

In order to be in correspondence with the redistribution model, the constant C_μ' has to be a function of the constants of the redistribution model. This means that C_μ' does not have to be equal to the C_μ from the k-ϵ turbulence model.

The dissipation tensor $\epsilon_{i,j}$ is commonly modelled using an isotropic model. This means that the dissipation is equal for all three normal stresses:

$$\epsilon_{ij} = \frac{2}{3}\tilde{\epsilon}\delta_{ij} \tag{4.25}$$

The values of the mostly used RSM model constant are given in Table 4.2.

The above described Reynolds stress and k-ϵ turbulence models were applied in predictions of the aerodynamics of a swirl burner. The obtained steady-state predictions were compared against cold flow measurements and discussed in detail in Section 6.4.1.

Recently, in research on classical turbulent combustion models, Reynolds-averaged Navier-Stokes (RANS) equations are being replaced by new unsteady approaches,

Table 4.2 The values of the constants in the RSM turbulence model.

C_1	C_2	$C_{\epsilon 1}$	$C_{\epsilon 2}$	C'_μ	σ_ϵ
1.8	0.6	1.44	1.92	0.065	1.3

large-eddy simulations (LES), because computers allow for the computation of the unsteady motions of the flames. The objective of LES is to explicitly compute the largest turbulence structure (eddies) of the flow field, whereas the effects of the smallest ones are modelled. This technique is currently widely used for nonreacting flows, but it is still at an early stage for combustion modelling [141] and at the very beginning of PF combustion modelling. Just a few attempts were made to apply LES methods to pulverised coal combustion in realistic complex geometries using a Lagrangian framework [142]. It is likely that the pace of developments will increase as computing resources become more powerful and as the CFD user community becomes more aware of the advantages of the LES approach for turbulence modelling.

In conclusion, in the field of turbulence modelling, there is no simple and general answer to the question, Which is the best model? Despite the advances, the goal of making reliable predictions of turbulence in practical problems has not yet been fully reached.

Similar to the Reynolds stresses, the turbulence scalar transport terms $\overline{\rho u_i'' Y_k''}$ and $\overline{\rho u_i'' h''}$ are generally closed, assuming gradient transport as follows:

$$-\overline{\rho u_i'' h''} = \frac{\mu_t}{\sigma_t} \frac{\partial \widetilde{h}}{\partial x_i} \tag{4.26}$$

$$-\overline{\rho u_i'' Y_k''} = \frac{\mu_t}{\sigma_Y} \frac{\partial \widetilde{Y}_k}{\partial x_i} \tag{4.27}$$

where μ_t is the turbulence viscosity provided by the turbulence model, σ_t is the turbulent Prandtl number, and σ_Y is the turbulent Schmidt number. In the present work, Eqs. 4.26 and 4.27 were employed with $\sigma_t = 0.7$, although other values can also be used.

It is common to neglect the gradient of the scalar-density correlations, $\Gamma_h \frac{\partial \widetilde{h''}}{\partial x_j}$ and $\Gamma_Y \frac{\partial \widetilde{Y_k''}}{\partial x_j}$ in Eqs. (4.3) and (4.4), respectively, and this practice has also been adopted in the present work. Thus the enthalpy Eq. (4.3) and chemical species Eq. (4.4) transport equations can be written in the following closed form:

$$\frac{\partial (\overline{\rho}\widetilde{h})}{\partial t} + \frac{\partial}{\partial x_j} \left(\overline{\rho}\widetilde{u}_j \widetilde{h} \right) = \frac{\partial}{\partial x_j} \left(\Gamma_{h,eff} \frac{\partial \widetilde{h}}{\partial x_j} \right) = \overline{\rho}\widetilde{S}_{ch} + S_h \tag{4.28}$$

$$\frac{\partial (\overline{\rho}\widetilde{Y}_k)}{\partial t} + \frac{\partial}{\partial x_j} \left(\overline{\rho}\widetilde{u}_j \widetilde{Y}_k \right) = \frac{\partial}{\partial x_j} \left(\Gamma_{Y,eff} \frac{\partial \widetilde{Y}_k}{\partial x_j} \right) = \overline{\rho}\widetilde{\omega}_k + S_{Y_k} \tag{4.29}$$

with $\Gamma_{h,eff}$ and $\Gamma_{Y,eff}$ being the effective diffusion coefficients that are the sum of the molecular and turbulence components.

The modelling of turbulent scalar transport terms requires some comments. The gradient transport assumption for turbulent scalar fluxes is directly inspired from turbulent scalar transport models in nonreacting, constant density flows. For such flows, Eqs. (4.26) and (4.27) may be derived by applying the so called Mischungsweg theory of Prandtl [143]. In the case of reactive turbulent flows, however, both experimental data [144] and theoretical analysis [145] have pointed out the existence of counter-gradient scalar turbulent transport (i.e., in the opposite direction than the one predicted using a gradient assumption) in certain flames. The counter-gradient diffusion occurs when the flow field near the flame is dominated by thermal expansion due to chemical reaction, whereas gradient diffusion occurs when the flow field near the flame is dominated by the turbulent motion [141]. The existence of counter-gradient turbulent scalar transport in some turbulent flames[3] (e.g., premixed flames) implies that simple algebraic closure based on gradient assumption and eddy viscosity concepts should be replaced by addition of transport equations for the turbulence fluxes in which unknown higher-order correlations appear. Such higher-order models are almost never found in commercial CFD codes because of their computational costs [141]. In fact, their complexity and the model uncertainties are too high to justify their implementation. Recent developments of LES in combustion provide an alternative for describing scalar turbulent transport by direct description at the resolved scale level without ad hoc modelling.

For high turbulence levels (as in most of the practical cases), however, the flame is not able to impose its own dynamics onto the flow field, and turbulent transport is expected to be of gradient type, as for any passive scalar. Therefore, this work uses this simplified approach.

Hybrid modelling of turbulent flows, combining RANS and LES techniques, has received increasing attention over the past decade to fill the gap between (U)RANS and LES computations in fluid dynamics applications at industrially relevant Reynolds numbers. With the advantage of hybrid RANS-LES modelling approaches, which are considerably more computationally efficient than full LES and more accurate than (U)RANS, particularly for unsteady fluid flows, has recently motivated numerous research and development activities.

4.1.2   Mean reaction rate

The modelling of gaseous chemical reactions is of particular importance since it concerns the complexity of the interaction between the turbulence and chemical kinetics. Each of these phenomena is extremely complex itself, making their mathematical modelling a very challenging task. In fact, chemical reactions such as those in combustion processes cannot be modelled entirely because of several physical preconditions, e.g., the large number of simultaneous reactions (i.e., 779 elementary reactions in case of methane oxidising under oxyfuel conditions [146,147]); the influence of the turbulence

---

[3]This phenomenon is generally explained by differential buoyancy effects induced by pressure gradients acting on heavy fresh gases and on light burnt gases. Thus, the turbulent diffusion can be depleted, which affects the dynamics of turbulent mixing [145].

on the chemical reaction rate; the influence of the chemistry on the turbulent flow, and so on.

Regardless of the flow type (laminar or turbulent), the reaction equations for a system with N_R chemical reactions with N species A_k, can be written as:

$$\sum_{k=1}^{N} v'_{kl} A_k \rightarrow \sum_{k=1}^{N} v''_{kl} A_k, \quad l = 1, \ldots N_R, \tag{4.30}$$

where v'_{kl} and v''_{kl} are stoichiometric coefficients for the species A_k in the reaction l.

The production rate (reaction rate) for species k, R_k $[kg/m^3 s]$, can be expressed as:

$$R_k = M_k \sum_{l=1}^{N_R} \left\{ (v''_{kl} - v'_{kl}) k_l \prod_{i=1}^{N} c_i^{v'_{il}} \right\} \tag{4.31}$$

where M_k is molar mass [kg/mol] for species k and c_i is the molar concentration [mol/m^3] of species i. The reaction coefficient k_l for reaction l can be given by the Arrhenius expression:

$$k_l = k_l(T) = A_l T^\beta exp\left(-\frac{E_{al}}{RT}\right) \tag{4.32}$$

Here, A_l is the pre-exponential factor, β is the temperature exponent, and E_{al} is the activation energy, all for reaction l.

The instantaneous chemical reaction rate is exponential in T and also depends non-linearly on the species concentrations; see Eqs. (4.31) and (4.32). The result of this nonlinearity is that the mean chemical reaction rate, represented by the term $\widetilde{R_k} = \overline{\rho} \widetilde{\omega}_k$ in the scalar equation representing species masses conservation Eq. (4.4), cannot be estimated as the instantaneous reaction rate using Favre-averaged quantities, and hence its approximation is extremely difficult. Therefore, reaction rate closures are generally developed from physical analysis, comparing typical time and length scales for turbulence and combustion to determine whether the combustion process is controlled by turbulent transport or chemical reactions.

Several possible approaches can be followed in order to overcome this difficulty. Although any attempt on classification will give "gray zones" between classes, the most commonly used models can be divided into:

1. Statistical models, which can be further derived into:
   - Moment closure methods, which are models with a series-expanded exponential Arrhenius expression and truncation of higher-order terms [148,149]
   - Turbulence mixing models, assuming that chemical reaction rates are mainly limited by turbulence mixing rates, e.g., the eddy breakup approach by Spalding [150] and the eddy dissipation concept by Magnussen [151,152]
   - Geometrical models, assuming the flame front as a geometrical surface evolving in the turbulence flow field—for instance, the "flamelet" hypothesis by Peters [153,154]

2. One-point statistical models, expressing the mean reaction rate by combining instantaneous reaction rates given from Arrhenius laws or chemistry tables with a multidimensional joint probability density function such as the PDF approach by Pope [155,156].

These models have been tested and have proved their applicability in various applications, from steady (based on conventional turbulence models) and, more recently, to unsteady (combined with large eddy simulations) reactive flows.

This work employs the eddy dissipation modelling approach for describing the mean reaction rate $\overline{\rho}\widetilde{\omega}_k$. Therefore, a more detailed description of this approach is given in the next section. The idea of representing the turbulent reaction rate as a function of turbulent dissipation rate is presented in several model variations. Earlier versions of this model are found in Magnussen and Hjertager [151] and Magnussen et al. [157]. Local extinction is modelled by Gran et al. [158]. Treatment of detailed chemical kinetics is shown by Gran and Magnussen [159] and Magel et al. [160]. The latest is found in Magnussen [152].

The eddy dissipation model (EDM)

Based on the idea proposed by Spalding [150] that the reaction rate will be determined by the mixing process or the rate of eddy breakdown, Magnussen and Hjertager proposed the eddy dissipation model (EDM) [151] that relates the reaction rate with the mean mass fraction of reacting chemical species:

$$\widetilde{\omega}_{EDM} = A\frac{\widetilde{\epsilon}}{k}min\left\{\widetilde{Y}_{fuel}, \frac{1}{r}\widetilde{Y}_{O_2}, B\frac{1}{1+r}\widetilde{Y}_{pr}\right\} \qquad (4.33)$$

where r is the stoichiometric oxygen requirement of the reaction and $\widetilde{Y}_{fuel}$ is the mean mass fraction of fuel (volatiles, Eq. (3.2), or CO, Eq. (3.3)); A and B are model constants that, compared to the k-ϵ turbulence model constants, can adopt different values for fitting purposes.

This formulation of mean reaction rate provides good predictions in many practical cases where fast chemistry assumption is justified [161]. However, the EDM fails in the prediction of flames, where the reactions in the near burner zone are kinetically controlled [162]. Therefore, an inclusion of chemical kinetics into the mean reaction rate formulation becomes necessary. Among different options, the series process approach, proposed by Azevedo [134], was considered in this study. It is based on the assumption that total reaction time can be represented as a sum of the characteristic time for turbulent mixing, $\tau_{EDM,k}$, and of the characteristic time for the chemical reaction $\tau_{kin,k}$:

$$\tau_{total} = \tau_{EDM,k} + \tau_{kin,k} \qquad (4.34)$$

Because the characteristic times are the reciprocal values of the respective reaction rates $\tau = 1/\omega$ for the mean reaction rate can be obtained:

$$\widetilde{\omega} = \frac{\widetilde{\omega}_{EDM,k}\widetilde{\omega}_{kin,k}}{\widetilde{\omega}_{EDM,k} + \widetilde{\omega}_{kin,k}} \qquad (4.35)$$

which in fact is the harmonic mean value.

The eddy dissipation concept (EDC) model

In contrast to EDM, the EDC model provides an empirical expression for the mean reaction rate based on the assumption that chemical reaction occurs in the regions where the dissipation of turbulence energy takes place. In flows of moderate to intensive turbulence, these areas are concentrated in isolated regions, occupying only a small fraction of the flow. These regions consist of "fine structures" whose characteristic dimensions are of the order of Kolmogorov's length scale in two dimensions, but not in the third [152]. The fine structures are not evenly distributed in time and space. They appear intermittently. Magnussen's model is based on the following main elements:

- A cascade model that describes the energy transfer from larger to smaller scales in turbulent flow
- An energy transfer model that expresses the characteristic quantities for the lowest level of scales as functions of scales from the large-scale level
- The large-scale levels are related to the mean flow by a turbulence model or resolved directly by LES
- The fine structures are assumed to be a steady-state homogeneous reactor; the chemical reactions occur there

In the EDC model, the fluid state is determined by the fine structure state, the surrounding state, and the fraction of fine structures. The ratio between the mass in the smaller eddies and the total mass in the surroundings is:

$$\gamma_\lambda = \left(\frac{3C_{D2}}{4C_{D1}^2}\right)^{1/4} \left(\frac{v.\tilde{\epsilon}}{\tilde{k^2}}\right)^{1/4} \tag{4.36}$$

where $C_{D1} = 0.134$ and $C_{D2} = 0.5$ are model constants [163]. Actually, γ_λ is a kind of intermittency factor, providing the possibility to find the fine structures or fine-structure regions in a given location.

The time scale for the mass transfer between the fine structures and the surroundings is estimated as:

$$\tau^* = \left(\frac{C_{D2}}{3}\right)^{1/2} \left(\frac{v*}{\tilde{\epsilon}}\right)^{1/2} = \frac{1}{\dot{m}^*} \tag{4.37}$$

where $\dot{m}^*$ is the mass flow over the boundaries of a fine structure.

Assuming all fine structures to be perfectly stirred reactors (PSRs), the mass exchange between the fine structures and the surroundings m^* can be linked to γ_λ^2/τ^*. This gives the following expression for the mean reaction rate:

$$\tilde{\omega}_k = \frac{\gamma_\lambda^2}{\tau^*} \left(Y_k^o - Y_k^*\right) \tag{4.38}$$

The mass averaged mean state can be found from the fine-structure state Y_k^* and the surrounding state Y_k^o as:

$$\tilde{Y} = \gamma_\lambda^2 Y^* + \left(1 - \gamma_\lambda^2\right) Y^o \tag{4.39}$$

Thus, the mean reaction rate can be represented as a function of the mean state values:

$$\widetilde{\omega}_k = \frac{\gamma_\lambda^2}{\tau^*} \left(\widetilde{Y}_k - Y_k^* \right) \tag{4.40}$$

where $\widetilde{Y}_k$ and Y_k^* are the mass fractions of species k as cell average and inside the fine structures, respectively.

The mass fractions inside the fine structure are determined by applying Arrhenius chemical reaction rates at fine structure conditions. This leads to a system of non-linear equations that are solved in an internal iterative procedure by applying different approaches [14,152,161,164]. In case of fast chemistry assumption, the kinetics inside the fine structures is not considered [152].

4.2   Coal particle modelling

In PF flame calculations, additional complexity is introduced by the two-phase character of the flow. To handle such a large variety of phenomena, two numerical approaches for particle-laden turbulent flows were developed: Euler-Eulerian and Euler-Lagrangian approaches [165]. In the framework of Eulerian formalism, the particles are assumed to behave as fluid, and so-called two fluid models were developed [165]. This method allows modelling of particle-particle stresses in dense particle flows using spatial gradients of particle volume fractions. However, the continuum assumption used in Eulerian particle models is not justified, because particles do not equilibrate with either local fluid or each other as they move through the flow field. Therefore, there is no physical justification to model particle dispersion by analogy to Fick's law (effective particle diffusivity). Moreover, when combustion is considered, the reaction rate depends on the history of each particle. Current Eulerian models are unable to account for particle history, because in the framework of the Eulerian approach the transport of the individual particle is not tracked [166]. In addition, modelling a distribution of types and sizes of particles complicates the continuum formulation because separate continuity and momentum equations must be solved for each size and type.

Using a continuum model for the fluid phase and a Lagrangian model for the particle phase allows an economical solution for flows with a wide range of particle types, sizes, shapes, and velocities [167]. The Lagrangian reference framework is the natural frame for treating combusting particles. The particles are treated as discrete objects and their motion is tracked as they move through flow field. It is possible to account for the noncontinuum behaviour of particles and particle history effects. The easiest way to compute a particle trajectory is to solve the equations of position and motion of a particle in a fluid. In such an approach, the Lagrangian particle equation for the mass center position $X_i^{(k)}$ of each k-particle can be written as:

$$\frac{dX_{p,i}^{(k)}}{dt} = u_{p,i}^{(k)} \tag{4.41}$$

For most practical dilute flow applications such as PF combustion, the *momentum equation* for a spherical particle moving in a viscous fluid and therefore experiencing a lift force due to the local gradients of translational fluid velocities can be greatly simplified by neglecting the Basset, virtual mass, Magnus, Saffmann, and buoyancy forces and therefore can be written as follows [166]:

$$\frac{du_{p,i}^{(k)}}{dt} = \frac{1}{\tau_p}\left(u_{f,i} - u_{p,i}^{(k)}\right) + g_i \tag{4.42}$$

where i represents the directions and τ_p is the particle relaxation time, defined as the rate of response of particle acceleration to the relative velocity between the particle and the bulk, expressed as:

$$\tau_p = \frac{24\rho_p d_p^2}{18\mu_f C_D Re_p} \tag{4.43}$$

where ρ_p, d_p, and Re_p are the particle density, diameter, and Reynolds number, respectively; μ_f is the fluid viscosity, and C_D is the drag coefficient, defined as [166]:

$$C_D = \left(\frac{24}{Re_p}\right)\left(1 + 0.15Re_p^{0.687}\right) \quad for\ Re_p \prec 10^3 \tag{4.44}$$

The main problem in solving Eq. 4.42 is to estimate the instantaneous fluid velocity at each particle location $u_{f,i}$ as the particle moves in discrete time steps through the Eulerian fluid velocity (Navier-Stokes) field. There are two main methods to define the fluid velocity: deterministic and stochastic. *Deterministic dispersion models* neglect the dispersive effect due to turbulence fluctuation. Thus the instantaneous fluid velocities are replaced by the time-averaged Eulerian fluid velocity, which is provided by the turbulence models. These types of models are easy to implement and do not require a significant number of trajectory calculations. However, due to obvious simplifying, they are unable to properly account for the dispersion phenomenon, which is of great importance in predicting the complex coal particle-turbulence interactions in the near burner zone of a PF swirl burner. Therefore, the use of *stochastic dispersion models* that consider the turbulence fluctuations is much more appropriate for PF combustion applications. In these models, the fluctuating fluid velocity that corresponds to a particular eddy is randomly sampled from a PDF that is obtained from the local fluid properties. The particle will interact with a particular eddy for a time, which is the momentum of the eddy lifetime and the eddy transit time. Knowing the eddy-particle interaction time and the randomly sampled fluctuating fluid velocity (which is assumed to be constant for that time), it becomes possible to solve the particle velocity according to Eq. (4.42).

During PF combustion, heat and mass transfer exist between particles and surrounding fluid. The *energy conservation equation of a spherical particle* in a reactive turbulent flow, assuming negligible internal temperature gradients, can be represented as:

$$\frac{d(m_p h_p)}{dt} = Q_{rad,p} - Q_{c,p} - r_p h_{p,g} \tag{4.45}$$

where $h_{p,g}$ is the enthalpy of the gas leaving the coal particle. The third term on the right side represents the enthalpy losses to the gas phase by transfer of mass due to particle reaction. This can be due to (1) water evapouration, considering constant particle temperature ($T_p = 100\,°C$) and $h_{p,g}$ to be equal to the enthalpy of evapouration; (2) devolatilisation, and/or (3) char burnout, with $h_{p,g} = h_{ch,g}$ being the energy released or absorbed during char surface reaction.

The net radiation to the particle can be calculated as a function of (1) the surrounding gas temperature T_w leading to $Q_{rad,p} = \sigma \epsilon_p A_p (T_p^4 - T_w^4)$, or (2) the irradiation G, calculated by the radiation model $Q_{rad,p} = \epsilon_p A_p (\sigma T_p^4 - G)$ [14]. $Q_{c,p} = h A_p (T_g - T_p)$ represents convection and conduction from the particle to the gas, with $h = \frac{Nu\,k_g}{d_p}$ being the heat transfer coefficient. The particle Nusselt number, Nu, is defined as $Nu = Nu_0 \frac{B}{e^B - 1}$, where Nu_0 is the correlation recommended by Crowe [126] and is given as $Nu_0 = 2 + 0.65\,Re^{1/2}\,Pr^{1/3}$ and the blowing parameter $B = \frac{c_p}{2\pi d_p k_g}\left(\frac{dm_p}{dt}\right)$ that considers the effects of high mass transfer on the convective heat transfer coefficient. The gas thermal conductivity k_g is calculated based on the local film temperature $(T_p + T_g)/2$. Together with the molecular viscosity μ, the heat conductivity k_g should consider the bulk atmosphere, in this particular case, air or O_2/CO_2.

The total mass of coal particle can be written as:

$$m_p = m_{H_2O} + m_{Vol} + m_{Char} + m_{Ash} \tag{4.46}$$

where m_p is the total particle mass, m_{Char} is the mass of char, m_{Vol} is the mass of volatiles, and m_{Ash} is the mass of ash in the coal. The variation of particle mass is related to the decrease in mass due to release of water vapour and volatiles and to the release of gaseous products formed during char conversion. Hence, the equation governing *the mass of coal particle during conversion* can be written as:

$$\frac{dm_p}{dt} = -\sum R_i \tag{4.47}$$

where R_i is the particle mass release rate [kg/s] due to particle drying, devolatilisation, and heterogeneous reactions between char and O_2, CO_2, H_2O, etc. The overall particle mass loss is modelled as a sum of the mass loses due to all heterogeneous reactions considered to happen simultaneously. Thus the predicted reaction rate will be somewhat higher than the real one because the reactants are actually competing for the same particle surface, as we discussed in Chapter 3.

The system of Eqs. (4.41), (4.42), (4.45), and (4.47) is solved for each particle trajectory (particle size and starting location) by integration over a time step small enough to minimise the errors.

The coupling between particles and the carrier gas phase models is normally achieved through definition of interphase exchange terms for mass, momentum, and energy sources using the particle-source-in-cell (PCE-CELL) algorithm, proposed by Crowe *et al.* [168]. Here particle properties are interpolated to the grid, and the flow field and particle interactions are calculated on the grid iteratively. Properties are then interpolated back to particles. The calculation of particle interactions on the grid increases computational efficiency with no significant increase in numerical error.

Andrews and O'Rourke [169] extended the PCE-CELL algorithm to a multiphase particle-in-cell (MP-PIC) method for one-dimensional Eulerian-Lagrangian flow. In this method, particles are treated both as particles and as a continuum. The particle stress gradient, which is difficult to calculate for each particle in dense flow, is calculated as a gradient on the grid and is then interpolated to discrete particles. Snider [167] extended the method to be two- and three-dimensional, with an improved grid-to-particle interpolation method. These methods consider the particle-particle interactions and are appropriate for modelling particle-dense flows such as in packed or fluidised beds. For PF modelling, however, the PCE-CELL approach remains still the most appropriate option, providing a good balance between accuracy and computational time requirements.

Because the numerical models for combustion processes as well as the numerical procedures are explained elsewhere [15,127–130,133,134,166], the focus of this chapter is on providing a more detailed look at the most recent existing models adopted for PF flames, with special attention given to oxyfuel conditions. The particle models discussed here were implemented in a specialised software developed by the author. The software is called Plug Flow Reactor (PFR) and validated against a wide range of experimental data for devolatilisation in Section 4.2.1 and for char conversion in Section 4.2.2 accordingly. The algorithm of the PFR program is shown schematically in Appendix B.1.

4.2.1   Devolatilisation models

As discussed in Section 3.2, coal particle devolatilisation is one of the most important processes in pulverised coal flames, and it strongly influences the ignition, flame stability, heat transfer, NO_x formation, char burnout, and so on. The process in itself represents the breakdown of the initial coal particle into a mixture of light gases and tars. For purposes of modelling the early stages of coal pyrolysis in a flame, a model is needed that can describe the effect of temperature on volatile yields as well as predict devolatilisation rates at high temperatures.

Two theoretical approaches for mathematical description of the pyrolysis can be considered. The first (empirical) method involves modelling the process using generalised expressions that rely on a reduced set of chemical reactions. The second (structural) method involves the use of a detailed model that accounts for the decomposition of the coal matrix during coal particle heatup and requires coal-specific analytical data to be modelled.

Empirical models

The empirical models are commonly employed within CFD codes whereby the coal particle devolatilisation process is generally described in terms of (1) a single-rate kinetic equation requiring three empirical parameters, namely, apparent activation energy E_1, apparent pre-exponential factor k_0, and ultimate yield V^*; or (2) by the two parallel reactions model, which requires input of six empirically derived parameters.

The *single overall reaction model* is based on the assumption that the rate of devolatilisation is proportional to the volatile matter that is to be released. The coal particle devolatilisation process is generally described in terms of a single-rate kinetic equation:

$$\frac{dV}{dt} = k\left(V^* - V\right) \tag{4.48}$$

with:

$$k = k_0 e^{-\frac{E}{RT_p}} \tag{4.49}$$

where V is the mass of volatiles and V^* denotes the asymptotic value of volatiles at $t \to \infty$ and can be calculated from $V_m Q$, where the factor Q is the ratio of weight loss to change in proximate volatile matter, in the range of 1.2 for coals with 80% carbon content and 1.7 for coals with 90% carbon content. V_m is the proximate volatile matter of coal on a dry ash-free (d.a.f.) basis. The fractional devolatilisation at any time is obtained after integration of Eq. 4.48, resulting in:

$$\frac{V}{V^*} = 1 - e^{-\int_0^t k \, dt} \tag{4.50}$$

Thus, the particle mass loss due to devolatilisation can be written as:

$$dm_{p,dev} = V^* C \left(1 - e^{-\int_0^t k \, dt}\right) \tag{4.51}$$

with V^*C being the fraction of the remaining volatile fraction to be released, with the function C defined as:

$$C = C e^{-\int_0^t k \, dt} \tag{4.52}$$

Some typical values for the kinetic parameters for single overall reaction models are given in Table 4.3. The parameters from Badzioch and Hawksley [170] models (1R-B) were obtained for temperatures up to 1273 K, at which the volatile yield is small, whereas the Goldberg and Essenhigh [171] model (1R-G) was derived for high temperatures and high heating rates, thus being more appropriate for PF flame applications.

The *two competing reactions model* represents the overall devolatilisation process by two mutually competing first-order reactions, where α_1 and α_2 are mass stoichiometric coefficients. The pseudo rate constants are given by

$$k_1 = k_{01} e^{-\frac{E_1}{RT_p}} \quad and \quad k_2 = k_{02} e^{-\frac{E_2}{RT_p}} \tag{4.53}$$

where k_{01} and k_{02} are pseudo pre-exponential factors and E_1 and E_2 are pseudo activation energies. According to Kobayashi et al. [172] (2R-K), the first reaction is assumed

Table 4.3 Kinetic parameters for the single overall reaction model.

	1R-B	1R-G
Activation energy, E (kJ/mol)	74	78.7
Pre-exponential factor, k_0 (1/s)	1.34×10^5[a]	5.5×10^6
Ultimate yield, V^*	$0.86 Q V_m$ ($Q = 1.41$)	$1.8 V_m$

[a] An average value for nonswelling coals, based on Badzioch and Hawksley's data [170], see Jamalludin et al. [173].

Table 4.4 Kinetic parameters for the two competing reactions model.

	2R-K	2R-U
Activation energy, E_1 (kJ/mol)	104.7	74
Activation energy, E_2 (kJ/mol)	167.5	251
Pre-exponential factor, k_{01} (1/s)	2×10^5	1.34×10^5[a]
Pre-exponential factor, k_{02} (1/s)	1.3×10^7	1.46×10^{13}
Ultimate yield, α_1	0.3	V_m
Ultimate yield, α_2	1.0	$2V_m^b$

[a] Adopted from 1R-B model.
[b] According to Jamalludin *et al.* [174].

to be dominant at relatively low temperatures, leading to the asymptotic volatile yield of α_1. At high temperatures the second reaction becomes faster than the first one, which requires that E_2 is much larger than E_1. This results in larger volatile yields. Thus, the overall particle mass loss can be expressed as:

$$dm_{p,dev} = \frac{m_{V_1} + m_{V_2}}{m_{coal,0}} = \int_0^t \left(\alpha_1 k_1 + \alpha_2 k_2\right) e^{-\int_0^t (k_1+k_2)dt} dt \qquad (4.54)$$

where $m_{coal,0}$ is the mass of the raw coal on a d.a.f. basis. Some typical values for the kinetic parameters are given in Table 4.4.

Both sets of parameters given in Table 4.4 are based on experiments at temperatures up to about 2100 K. The reaction rate of the first reaction in the 2R-U model was determined using the same pre-exponential factor and the same activation energy as in the 1R-B model, following the suggestions of Jamalludin *et al.* [174]. Thus, the modified model of Ubhayakar *et al.* [175] (2R-U) predicts experimental data reasonably well, especially at higher heating rates (above 10^5 K/s), as shown in Figure 4.3.

Structural models

Recently, significant progress has been made in modelling coal devolatilisation considering coal chemical structure and the related reaction mechanisms. Coal devolatilisation was modelled viewing coal as a macromolecular network structure. Three types of network models were developed that consider the thermal decomposition of coal by the breakdown of the coal macromolecular network. These models are the FLASHCHAIN model proposed by Niksa [176]; the function group, depolymerisation, vaporisation, and cross-linking (FG-DVC) model proposed by Solomon *et al.* [177]; and the chemical percolation devolatilisation (CPD) model proposed by Fletcher and co-workers, as summarised and compared in [15,64]. These three models, based on coal chemical structure parameters, are capable of predicting the devolatilisation process, volatile species yields, and the structure of coal under various devolatilisation conditions (pressure, temperature, heating rates) and for a wide range of coal types. In the present work,

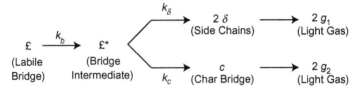

Figure 4.1 Scheme of the reaction paths considered in CPD model [64].

the CPD model [40,64,178] was employed to describe the coal devolatilisation process. In the CPD model, the coal structure is defined with four chemical structural parameters, i.e., the initial fraction of intact bridges in coal, p_0; the coordination number (number of attachments per cluster), $\sigma + 1$; the average molecular weight per aromatic cluster, M_{clust}; and the average molecular weight per side chain, M_δ. These parameters were generally taken directly from ^{13}C NMR (nuclear magnetic resonance) measurements [40,64]. The simple reaction sequence proposed for coals [40,178] starts with (1) the breaking of a chemical bond in a labile bridge to form a highly reactive bridge interme-diate (e.g., two free radical side chains temporarily trapped in the reaction cage), which is rapidly consumed by one of two competitive processes. The reactive bridge material (2) may be released as light gas, with the concurrent relinking of the two associated sites within the reaction cage to give a stable or charred bridge, or (3) may be stabilised (e.g., due to hydrogen extraction by free radicals) to produce side chains from the reactive bridge fragments. The stabilised side chains may be (4) converted eventually into light gas fragments through a subsequent, slower reaction. Thus, the simple scheme shown in Figure 4.1 was proposed to represent the devolatilisation process.

A labile bridge, represented by £, decomposes by a relatively slow step with rate constant k_b to form a reactive bridge intermediate, £*, which is unstable and reacts quickly in one of two competitive reactions. In one reaction pathway, the £* is cleaved with rate constant k_δ, and the two halves form side chains δ that remain attached to the respective aromatic clusters. The side chains eventually undergo a cracking reaction to form light gas. In a competing reaction pathway, the £* is stabilised to form a stable "char bridge" c with associated release of light gas (rate constant k_c). The competitive processes depend only on their reaction rate, defined with a composite rate coefficient $\rho_{CPD} = k/k_c$. For coals, ρ was taken as 0.9 [178], meaning that the two pathways almost have the same importance. Percolation lattice statistics were employed to describe the generation of tar precursors (metaplast) based on the number of cleaved labile bridges in an infinite coal lattice. The tar release was described basically as a competitive process between evapouration and cross-linking. The coal-dependent chemical structure parameters make the model applicable to a wide range of coals and reaction conditions.

Devolatilisation rates

The models we have discussed were validated against measurements of as-fired frac-tions (75% less than 75 μm) of bituminous (Pittsburgh) and subbituminous (Blue)

Table 4.5 Coal properties and chemical structure parameters for CPD input data.

Seam Name Coal Rank	Pittsburgh #8 hv Bituminous	Blue #1 Subbituminous
Proximate analysis, (wt-%, as received)		
Volatile	33.56	40.33
Fixed carbon	50.58	43.41
Ash	13.32	4.04
Moisture	2.54	12.22
Ultimate analysis, (wt%-d.a.f.)		
Carbon	83.26	74.97
Hydrogen	5.41	5.64
Nitrogen	1.58	1.4
Oxygen	8.1	17.20
Sulphur (+ chlorine)	1.67	0.78
Coal structure parameters		
p_0	0.62	0.42
$\sigma + 1$	4.5	5.0
M_{clust}	294	410
M_δ	24	47
c_0	0	0.15

coals, devolatilising in a transparent-wall flow reactor of the Sandia Coal Devolatilisation Laboratory (CDL), described in detail by Fletcher et al. [179]. For these coals, the chemical structure parameters, which are necessary inputs for the CPD model, were calculated with the empirical correlation based on elemental composition and volatile matter content proposed by Fletcher et al. [64]. The parameters used are listed in Table 4.5, together with the CDL analysis results.

The parameters used for the empirical models are the same as listed in Tables 4.3 and 4.4. All the cases consider changes in the particle mass (resp. density) at constant particle diameter, thus neglecting "swelling" effects. Two reactor temperature conditions were considered: 1050 K and 1250 K. The particles enter the reactor at room temperature. The predicted particle temperatures are compared with the measured ones as shown in Figure 4.2.

The rates of mass release obtained when applying five different devolatilisation models are plotted in Figure 4.3. Although the volatile yield depends on many parameters, as discussed in Section 3.2, the volatiles' release here was artificially restrained up to the ultimate analysis volatile content. This was done in order to keep the mass and energy balance closed when the model was used as UDF for the Fluent CFD solver.

As shown, the predictions obtained by applying the CPD model as well as the two-reaction model by Ubhayakhar et al. (2R-U) and by the one-reaction model of Badzioch and Hawksley [170] (1R-B) are in good agreement with experimental data

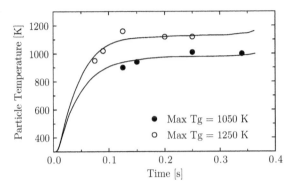

Figure 4.2 Comparison of measured (discrete points) and calculated (curves) particle temperatures as a function of residence time for two temperatures in the CDL reactor.

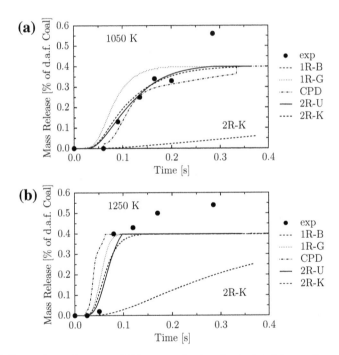

Figure 4.3 Comparison of measured (discrete points) and calculated (curves) mass release as a function of residence time for bituminous coal particles (70 μm size fraction) using different models for (a) 1050 K gas conditions and (b) 1250 K gas conditions.

for a reactor temperature of 1050 K. The 1R-G model slightly over-predicts the volatile release (due to its higher pre-exponential coefficient), whereas the two-reaction model using parameters proposed by Kobayashi *et al.* (2R-K) fails in predicting the volatile release. Similar trends are obtained for 1250 K reactor temperature. The 2R-K model fails in predicting the mass release, whereas the CPD leads to a slight over-prediction of the rates. The rest of the models adequately predict the release of volatile matter under the given conditions. The upper limit of volatile matter release was numerically limited to the value from the proximate analysis, thus resulting in a nonphysical straight horizontal line in the figures.

Based on the results discussed above, an additional test run was made with the aim of testing the model prediction behaviour on devolatilisation rates of subbituminous coal (Blue). Here just CPD and the two-reaction model based on coefficients provided by Ubhayakar *et al.* were implemented. The results, plotted in Figure 4.4, show an over-prediction of devolatilisation rates by the CPD model for both temperature conditions. In comparison, the two-reaction model predicts the pyrolysis of subbituminous coal as well as that of bituminous coal, without the need to adjust the three empirical parameters.

4.2.2   Char burnout models

The main heterogeneous reactions in a pulverised fuel flame occur with char and oxygen, Eq. (3.4); with char and CO_2, Eq. (3.5); and with char and water vapour, Eq. (3.6). For modelling purposes, these reactions are normally approximated or simplified as the reaction of carbon with oxygen and those of carbon with CO_2 and with water vapour.

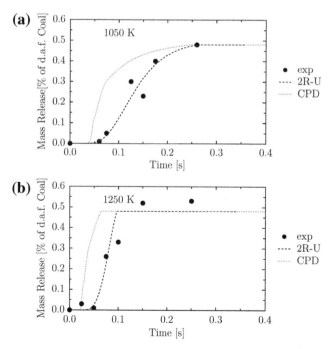

Figure 4.4 Comparison of measured (discrete points) and calculated (curves) mass release as a function of residence time for subbituminous coal particles (70 μm size fraction) using different models for (a) 1050 K gas conditions and (b) 1250 K gas conditions.

To formulate a simple model for the combustion characteristics of a single char particle, the factors that need to be considered are the intrinsic reactivity of the char, the total surface area (including pore surface area), the partial pressure of oxidant(s), and the temperature of the particle and the surrounding gas. Commonly, char reaction rate is expressed as follows:

1. Empirically as an Arrhenius formulation based on one-step heterogeneous reaction assumption; or
2. By a Langmuir model based on a two-step (adsorption and desorption) mechanism

Kinetics with a one-step mechanism can be further classified into global or *apparent* kinetics and *intrinsic* kinetics [180]. The apparent kinetics consider an overall reaction rate and base the reaction rate on the external surface area and on the oxygen concentration at the particle surface, whereas the intrinsic approach considers the particle reaction rate as a function of the intrinsic reactivity, pore surface area, and local oxygen (or CO_2, H_2O) concentrations. Using either apparent kinetics or intrinsic kinetics, three kinetic parameters are commonly used: activation energy (apparent activation energy E_a or intrinsic activation energy E_i), reaction order (apparent order n or intrinsic order m), and pre-exponential factor. It is well known that these parameters, generally obtained from experiments, vary with char properties and reaction conditions. Even so, similarity of the char chemical structure [181] implies the similarity in char reactivity, thus implying similar activation energy [180]. Therefore the char reactivity can be described in a more general kinetic expression, particularly for high-temperature combustion [70,182]. In the scope of the present study, a char combustion model approach was developed that includes these two kinds of kinetic approximations. For this, the activation energy and the reaction order were kept constant while, for convenience, the pre-exponential factor was left adjustable to relate to the fuel dependence of char reactivity and reaction conditions. Alternatively, a simple linear dependence of the reaction rate on the conversion ratio [183] was included and tested. Consequently, the dependence of the char reaction model on empirically derived parameters was considerably reduced, although not completely eliminated.

Langmuir kinetics consider the carbon-oxygen (or $C-CO_2$; $C-H_2O$) reaction involving chemisorption and desorption processes [57,184]. This model attempts to describe char reaction on a more fundamental basis and eliminates the need to fix a reaction order. Literature results indicate that the kinetic parameters still need to be improved and generalised so that the model can be used for engineering purposes. Here a linear dependence of the reaction rate on the conversion ratio [183] was also included to generalise the kinetics.

Routines were prepared to simulate char combustion following each of the three kinetic mechanisms, identified as shown in Appendix A.1. All models include in detail the heat transfer between the particle and environment, the external boundary mass diffusion, and chemical reactivity. For the intrinsic reactivity and Langmuir kinetics, inter-particle diffusion is also considered through effectiveness factors. All the models were validated with the char combustion results obtained from isothermal plug flow reactor (IPFR) experiments. These models for char reaction rate, together with the respective kinetics, were coupled with the particle devolatilisation and particle drying models and together applied later to a CFD simulation of PF oxy-firing burners. The coupling algorithm is based on the assumption that the volatile release and the char conversion can happen simultaneously. This means that the char ignition can start before the end of the devolatilisation process. Details concerning the impact of this assumption on the coal particle conversion rate can be found in Schiebahn [88].

Char-O_2 reaction models

The oxidation of carbon can be presented in a general form as:

$$C + \frac{1}{\phi}O_2 \rightarrow \left(2 - \frac{2}{\phi}\right)CO + \left(\frac{2}{\phi} - 1\right)CO_2 \tag{4.55}$$

where $\phi = \frac{2p+2}{p+2}$ is the oxidiser-to-fuel ratio, indicating the composition of the reaction products, and can take values between 1 (with CO_2 being the only product) and 2 (with CO being the only product). The formation of any of these products is often empirically correlated with an Arrhenius equation such as:

$$p = \frac{[CO]}{[CO_2]} = A_0 e^{-B_0/T_p} \tag{4.56}$$

with the coefficients suggested by Arthur [185]: $A_0 = 2500$, and $B_0 = 6240$ K for a particle diameter less than 50 μm. For bigger particles $(D_p \succ 1\ mm)\ \phi = 1$ and for intermediate diameters, a linear interpolation can be made.

The molar ratio between CO and CO_2 can be also expressed as:

$$\frac{[CO_2]}{[CO]} = A_0 p_{O_2,s}^{\eta_0} e^{-B_0/T_p} \tag{4.57}$$

with the coefficients suggested by Tognotti et al. [186]: $A_0 = 0.02$, $B_0 = 3070$ K, and $\eta_0 = 0.21$. Then $\phi = \frac{1+\psi}{2}$, where $\psi = CO_2/CO/(1 + CO_2/CO)$.

Generally, the stoichiometric coefficient $\nu_O = \frac{1}{\phi}$ of oxygen represents a major uncertainty in modelling char oxidation.

Boundary layer diffusion

Taking oxygen as conserved in the boundary film leads to the simplified equation for the reaction rate controlled by the boundary film diffusion:

$$R_{char,O_2} = k_D \left(P_{O_2,\infty} - P_{O_2,s}\right) \tag{4.58}$$

where $P_{O_2,\infty}$ and $P_{O_2,s}$ are the partial pressures of oxygen in the bulk and at the particle surface, respectively; k_D is the boundary mass transfer coefficient, obtained from Sherwood number correlations and defined as:

$$k_D = \frac{Sh\ D_{O_2,m}\phi M_C}{d_p R T_m} \tag{4.59}$$

with Sherwood number $Sh = 2$ for pulverised char particles [57], $M_C = 12$ the molecular weight of carbon, $T_m = (T_{gas} + T_s)/2$ the mean temperature of the boundary layer, d_p the particle diameter, and $D_{O_2,m}$ the binary diffusion coefficient for oxygen calculated according to VDI-Warmeatlas [33].

At high temperatures, surface reaction is so fast that the surface oxygen concentration approaches zero and the overall reaction rate approaches the maximum value allowed by boundary layer diffusion:

$$R_{char,O_2,max} = R_{diff,P_s=0} = k_D P_{O_2,\infty} \tag{4.60}$$

In this case, the overall reaction rate is solely controlled by boundary layer diffusion corresponding to Zone III combustion conditions.

Apparent kinetics

Char reaction rate per particle external surface area is represented as an *n-th* order rate equation as follows:

$$R_{char,a} = k_a P_s^n \tag{4.61}$$

Here k_a is the apparent reaction constant, represented as an Arrhenius expression $k_a = A_a e^{\frac{-E_a}{RT_p}}$. Using the apparent kinetics, the thermal annealing and ash inhibition were included, considering the reactivity loss at the late stage of char combustion following Hurt *et al.* in their CBK model [70]. Two reaction orders were considered: first reaction order ($n = 1$), based on Field's data [81] with an activation energy of 18 kcal/mole, and half reaction order ($n = 0.5$) according to Hurt *et al.* [70], with the activation energy taken as 20 kcal/mol, whereas the pre-exponential factor was adjusted and fitted to the experimental data (see Table 4.6).

Intrinsic kinetics

For the intrinsic kinetics, the reaction rate per external surface area can be expressed as [187]:

$$R_{char,O_2} = \eta \gamma \rho_p A_g k_i P_s^m \tag{4.62}$$

with an intrinsic rate constant, $k_i = A_i e^{\frac{-E_i}{RT_p}}$. In this case, the single-pore model [187] was employed, whereas the pore surface area available in an oxidation reaction is

Table 4.6 Kinetic parameters for the apparent reaction model.

	First Order[a]	Half Order (CBK)[b]
Activation energy, E_1 (kJ/mol)	74	78
Pre-exponential factor, k_{01} (kg/m^2-s-Pa^n)	0.005	$10.*\exp(10.82 - 0.0714 C_{coal\%,daf})/318$
Reaction order, n	1.0	0.5

[a] For nonswelling bituminous coals, according to Smith [187].

[b] Here $C_{coal\%,daf}$ means the carbon content (in %) on a d.a.f. basis.

denoted with an empirical effectiveness factor, η, which is a function of the Thiele modulus [180]:

$$\eta = \frac{3}{Thiele} \left(\frac{1}{tanh(Thiele)} - \frac{1}{Thiele} \right)^{0.5} \tag{4.63}$$

For m-th order kinetics, the Thiele modulus can be written as:

$$Thiele = \gamma \left(\frac{A_g \rho_p k_i p_s^{m-1}}{D_e} \right)^{0.5} \left(\frac{RT_p}{12\vartheta} \right)^{0.5} \tag{4.64}$$

where the effective pore diffusion coefficient D_e incorporates the effect of the number and size of the pores in the unit cross-section and the tortuosity (τ) of these pores; $\gamma = \frac{d_p}{6}$ is the characteristic size of the particle; A_g is the particle's specific area, and m is the true order of reaction.

Although three values of intrinsic activation energy are typically reported in literature, i.e., 42.8 kcal/mol [187], 38.6 kcal/mol [188], and 35 kcal/mol [182], the first value (proposed by Smith [187] for bituminous coals) was used in this study. The activation energy and reaction order (m = 1) were fixed, whereas the pre-exponential factor was left adjustable.

Langmuir-Hinshelwood kinetics

The extended resistance equation considering an adsorption/desorption two-step Langmuir reaction mechanism developed by Essenhigh [184] was used to calculate the char reaction rate. Combining this with the external diffusion, an extended reaction rate can be written as:

$$\frac{1}{R_{\text{char},i}} = \frac{1}{k_D p_g} + \frac{1}{\epsilon k_{ad} p_g} + \frac{1}{\epsilon k_{de}} \tag{4.65}$$

where k_{ad} and k_{de}, both having an Arrhenius form, are the rate constants for the adsorption and desorption reactions, respectively. The second effectiveness factor, ϵ, was used to consider the effect of interparticle diffusion. Four kinetic parameters are required: the activation energy of adsorption and desorption reactions, E_{ad} and E_{de}, and their pre-exponential factors, k_a and k_d. The values of these parameters have been reported in the literature. In the review of Laurendeau [57], the activation energies of E_{ad} and E_{de} from several researchers were located in the ranges of 1–23 kcal/mol and 40–80 kcal/mol, respectively. In the present work, the activation energies amount to 10.04 kcal/mol for adsorption and 32.95 kcal/mol for desorption, whereas the pre-exponential constant of desorption is 26170 kg/m^2-s, following Essenhigh [59] and after parameter comparison. The pre-exponential factor of adsorption was left adjustable to fit reaction conditions. The second (Thiele) effectiveness factor ϵ, employed by Essenhigh, was calculated from the power index of the normalised density-diameter relationship [184]. The power index is very difficult to determine accurately, since it varies over several orders of magnitude. For example, Essenhigh [184] reported a value of around 1 for high-temperature char oxidation and values in the range of 104–105 for low-temperature oxidation. In the present work, a value of $\epsilon = 3.6$ was considered following Essenhigh and Mescher [189].

Global reaction rates

The char reaction rate as well as particle temperature can be calculated by solving the coupled equations for boundary diffusion, Eq. (4.58); kinetic equation, Eqs. 4.61, 4.62, or 4.65; and the energy balance equation for a char particle.

In case of PF flames, the char-O_2 reaction occurs under Zone II conditions or in the transition region between Zone I and II [15]. Therefore, both oxygen diffusion and reaction kinetics have to be considered. The processes of diffusion of reactants through the boundary layer and of heterogeneous reaction on the char particle surface can be assumed to occur in series, and thus the global reaction can be written as follows:

- For apparent and intrinsic reaction rates of the first order:

$$R_{char,global} = \left(\frac{1}{k_D} + \frac{1}{k} \right)^{-1} P_{O_2,\infty} \tag{4.66}$$

with k being k_a or $\eta \gamma \rho_p A_g k_i$ for apparent and intrinsic kinetics, respectively.
- For apparent reaction rate of the half order:

$$R_{char,global} = k_a \sqrt{P_{O_2,s}} = \frac{-k_a + \sqrt{k_a^2 + 4k_D^2 P_{O_2,\infty}}}{2k_D} \tag{4.67}$$

Supplemental mechanisms An ash inhibition submodel was included following Hurt et al. [70] to consider the influence of the ash film on oxygen diffusion. For one-step reaction kinetics, the thermal annealing submodel from Hurt et al. [70] was also included to consider the variation of the char reaction under heating. An alternative approach is to assume a linear decrease of the chemical reaction rate during burnout, according to Moors et al. [190]:

$$k_{a,corrected} = (1 - BO) k_a \tag{4.68}$$

with BO being the char burnout, defined as:

$$BO = \frac{M_{p0} (CoalChar - CoalAsh) - M_p}{M_{p0}CoalChar} \tag{4.69}$$

with the initial mass of coal particle M_{p0}; current mass of coal particle M_p; and CoalChar and CoalAsh the initial mass fractions of char and ash in a coal particle, respectively.

The mode of particle burning can be considered following Hurt et al. [70] to describe the evolution of the particle diameter and density along the combustion process. The apparent density of the coal char particle is expressed by the following correlation:

$$\frac{1}{\rho_p} = \frac{x_a}{\rho_a} + \frac{(1 - x_a)}{\rho_c} \tag{4.70}$$

where $x_a = CoalAsh$ is the mass fraction of ash in the particle, and ρ_a and ρ_c are the apparent densities of ash and combustibles, respectively. The variation of apparent density of a combustible is related to burnout by

$$\frac{\rho_c}{\rho_{c,0}} = \left(\frac{M_c}{M_{c,0}}\right)^{\alpha} \tag{4.71}$$

where α is the empirical burning mode parameter [70,191]; M_c is the current carbon content in a char particle; and $M_{c,0} = M_{p0}CoalChar$ is the initial carbon content in the char particle. After calculating a particle density, the particle size can be calculated as:

$$\frac{d_p}{d_{p,0}} = \left[\left(\frac{M_p}{M_{p,0}}\right)\left(\frac{\rho_{p,0}}{\rho_p}\right)\right]^{1/3} \tag{4.72}$$

As particle size and density are known, the particle porosity, needed when using intrinsic kinetics, can be calculated accordingly.

Char-O$_2$ reaction model validation

The char-O$_2$ reaction models were validated with the experimental results obtained for three different coal chars under conditions similar to an industrial pulverised coal flame; see Table 4.7. The residence time in these reactors, operating under atmospheric pressure, is 1.5 s, so it is possible to measure burnout behaviour and apparent reaction rate of particles of commercial sizes. High flue gas inlet temperatures guarantee heating rates typical for PF combustion. The Gottelborn and Kellingley char were produced by devolatilising the as-fired samples of Gottelborn and Kellingley bituminous coal at 1673 °C in an oxygen-free environment in the same reactor. Then the chars were burnt in an IPFR at IFRF. The experiments with Gottelborn char were conducted at 5 vol% oxygen (rest nitrogen) and at reactor temperatures of 1233 K, 1473 K, and 1673 K, respectively [183]. The experiments with Kellingley char were conducted at 4, 9, and 12 vol% oxygen (rest nitrogen) and at reactor temperatures of 1223 K, 1473 K, and 1673 K

Table 4.7 Proximate and ultimate analysis for investigated coals.

	Gottelborn[183]	Kellingley[192]	Low Rank[193]
Proximate analysis (wt%)			
Volatile (dry)	36.9	32.1	42.8
Fixed carbon(dry)	54.0	52.4	56.1
Ash (dry)	9.1	15.4	1.1
Moisture	1.7	2.9	10
Ultimate analysis (wt%-d.a.f.)			
Carbon	72.2	82.9	
Hydrogen	4.8	5.4	
Nitrogen	1.5	2.2	
Oxygen	11.0	7.4	
Sulphur (+ chlorine)	1.1	2.9	

[192], respectively. Additionally, the carbon burnout of low-rank coal was investigated in a drop tube furnace, described elsewhere [193]. In this case, coal particles with narrow-size fractions between 106 and 300 μm were burnt in 14 vol% oxygen at 1173 K.

The inputs for the char burnout model include reaction conditions (furnace temperature and oxygen concentration) and char properties (particle size distribution, ash content, particle density, and kinetic parameters). The related data, together with kinetic parameters for the three fractions, are summarised in Table 4.8. The model calculations are compared with experimental data, shown in Figures 4.5–4.8.

The prediction of the CBK char burnout model was compared with the experimental results for Gottelborn coal char, shown in Figure 4.5 as a function of reactor temperature and in Figure 4.6 as a function of reactor temperature and oxygen concentration. It can be found that when the ash inhibition and/or thermal annealing are not included, the model overestimates the combustion at the later stages, justifying their importance (not shown here). Additionally, the pre-exponential factor A_C must be varied for lower

Table 4.8 Kinetic parameters for char burnout models for three different coal chars.

Seam Name Coal Rank	Gottelborn Bituminous	Kellingley Bituminous	hv. Bituminous
Density (kg/m^3)	900	992	990
Porosity	0.5	0.39	–
Diameter (μm)	80	80	Varies
Temperature (K)	Varies	Varies	1173
Oxygen conc. (vol%.)	5	Varies	14
Apparent Kinetics			
Pre-exponential factor $(kg/m^2 s Pa)(n=1)$	0.005	0.005	0.005
Pre-exponential factor $(kg/m^2 s Pa^{0.5})(n=0.5)$			
without TA	9.1[a] 0.2 (all temp)	4.22	
with TA	0.6[a] 0.7(1233 K, 1473 K) 0.9(1673 K)	0.28[a] 1.4(1673 K, all O_2,%)	
Intrinsic Kinetics			
Pre-exponential factor $(kg/m^2 sPa)$	0.16 (1673 K) 0.6 (1473 K) 4.6 (1223 K)	4.8 (1673 K, all O_2,%) 4.8 (1223 K, 4% O_2) 5.8 (1223 K, 5.8% O_2) 7.8 (1223 K, 12% O_2)	
Langmuir Kinetics			
(k_{ad})	100 (1673 K) 40 (1473 K) 4 (1223 K)	100 (1673 K,all O_2,%) 2.8 (1223 K, 4% O_2) 5 (1223 K, 9% O_2) 30 (1223 K, 12% O_2)	

[a] Calculated value with empirical correlation related to coal carbon content [70,182].

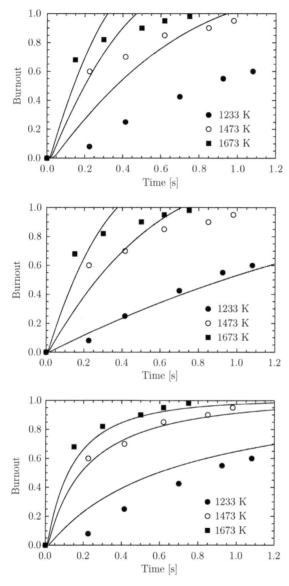

Figure 4.5 Char-O_2: Comparison of measured (discrete points) and calculated (curves) mass release as a function of residence time for bituminous Gottelborn coal char particles using (top) the CBK model, (middle) the CBK model with adapted Ac, and (bottom) the CBK model with BO.

temperature combustion (see Table 4.8). The model can only represent the burnout process at early stages but overestimates at the late stage of char oxidation, when the thermal annealing and the ash inhibition effects gain importance. Additionally, the pre-exponential factor must be adapted to fit different reaction temperatures (see

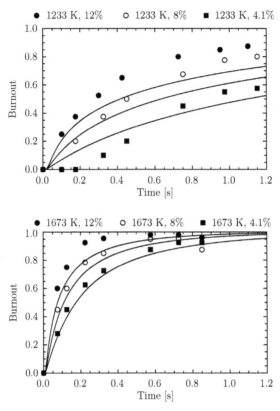

Figure 4.6 Char-O_2: Comparison of measured (discrete points) and calculated (curves) mass release as a function of residence time for various oxygen concentration for bituminous (Kellingley) coal char particles using the CBK model with BO at (top) 1233 K, and (bottom) 1673 K.

Figure 4.5, middle). However, the pre-exponential factors used here are significantly lower than those given in the empirical correlation by Hurt *et al.* [70]. Inclusion of a linear dependence of the burnout progress on the reaction rate (BO model) leads to a significant improvement in the model predictions (see Figure 4.5, bottom). This model combination (CBK with BO) also adequately predicts the influence of variations in oxygen concentration (see Figure 4.6).

The prediction of the char burnout model with *intrinsic kinetics* using kinetic parameters originally proposed by Smith [187] overestimates the char burnout rates at the late stage, especially for high temperatures (not shown here). Therefore, an adaptation of the pre-exponential factor A_c is required as shown in Figure 4.7 (top) or a combination with the BO submodel, Figure 4.7 (bottom). This model combination (intrinsic kinetics with BO) also adequately predicts the influence of variations in oxygen concentration, as shown in Figure 4.8.

The prediction of the *L-H* char burnout model using kinetic parameters that was originally proposed by Essenhigh [189] overestimates at the late stage the char burnout rates, especially for higher temperatures, as shown in Figure 4.9 (top). Therefore,

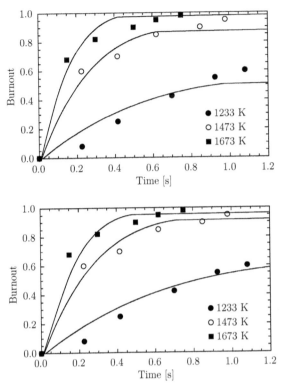

Figure 4.7 Char-O_2: Comparison of measured (discrete points) and calculated (curves) mass release as a function of residence time for bituminous Gottelborn coal char particles using (top) the intrinsic model with adapted Ac and (bottom) the intrinsic model with BO.

combination with the BO submodel is justified, and as shown in Figure 4.9 (bottom), this leads to a significant improvement in the predictions. This model combination (L-H kinetics with BO) also adequately predicts the influence of variations in oxygen concentration (see Figure 4.10).

The prediction of the *first-order apparent kinetics* char burnout model using kinetic parameters originally proposed by Field [81] overestimates at the late stage the char burnout rates, especially for higher temperatures (not shown here). Therefore, a combination with the BO submodel without changing the kinetic parameters is justified, and as shown in Figure 4.11, this leads to an significant improvement in the predictions. This model combination (first-order apparent kinetics with BO) also adequately predicts the influence of variations in particle diameter (see Figure 4.12). This model setup was used for the burner simulations discussed in Chapter 6.

Char-CO_2 reaction models

Applying an *n-th* order apparent reaction rate equation, Eq. 4.73, the overall char-CO_2 gasification rate per-particle external surface area is assumed to be proportional to the

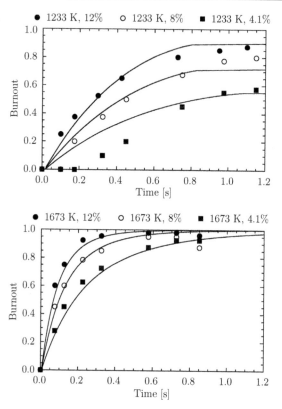

Figure 4.8 Char-O_2: Comparison of measured (discrete points) and calculated (curves) mass release as a function of residence time for various oxygen concentrations for bituminous (Kellingley) coal char particles using the intrinsic kinetics model with BO at (top) 1233 K and (bottom) 1673 K.

n-th power of partial pressure $P_{CO_2}^n$ of the gaseous reactant (in this case, CO_2):

$$R_{char,CO_2} = k_a P_{CO_2}^n \tag{4.73}$$

where n is the reaction order and k_a is the apparent reaction constant, represented as an Arrhenius expression $k_a = A_a e^{\frac{-E_a}{RT_p}}$.

Assuming intrinsic kinetics, the reaction rate per external surface area can be expressed according to Smith [187] as:

$$R_{char,CO_2} = \eta S k_i P_{CO_2}^m \tag{4.74}$$

with the intrinsic rate constant, $k_i = A_i e^{\frac{-E_i}{RT_p}}$. The internal surface area S depends on the coal type and changes with the conversion ratio x.[4]

$$x = \frac{M_0 - M(t)}{M_0 - M_{ash}} \tag{4.75}$$

---

[4]Here the conversion ratio considers only the char particle, not the whole coal particle, which usually contains all combustible matter.

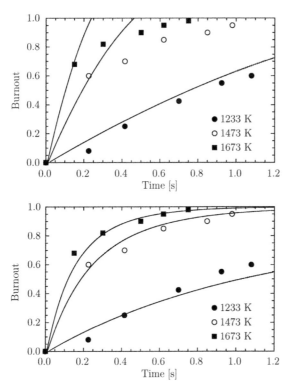

Figure 4.9 Char-O_2: Comparison of measured (discrete points) and calculated (curves) mass release as a function of residence time for bituminous (Gottelborn) coal char particles using (top) the L-H model and (bottom) the L-H model with BO.

where M_0 is the initial particle mass, $M(t)$ the current particle mass, and M_{ash} the mass of ash in the particle. By applying a random pore model, S becomes:

$$S = S_0 (1 - x) \sqrt{1 - \Psi ln (1 - x)} \tag{4.76}$$

where the initial internal surface area is S_0 for $x = 0$, and Ψ is a structural parameter that considers the nonlinear dependence between the conversion rate and the conversion ratio (see Appendix A.1.1). The pore surface area available in gasification reactions is denoted, similar to the char-O_2 reaction, with an empirical effectiveness factor η, which is a function of the Thiele modulus. Table 4.9 summarises the parameters fitted for n-th order reaction kinetics for nine different types of chars.

The extended resistance equation, which considers an adsorption/desorption two-step Langmuir reaction mechanism, Eqs. (3.7)–(3.8), and steady state for C(O), is used to calculate the char-CO_2 reaction rate. Combined with the external diffusion, an extended reaction rate can be written as:

$$R_{CO_2} = S \frac{k_{1f} P_{CO_2}}{1 + \frac{k_{1f}}{k_3} P_{CO_2} + \frac{k_{1b}}{k_3} P_{CO}} \tag{4.77}$$

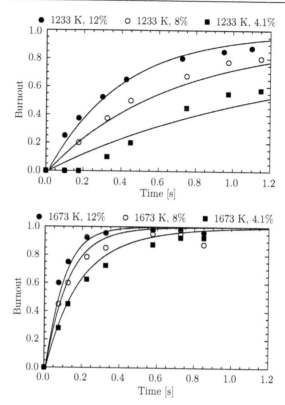

Figure 4.10 Char-O_2: Comparison of measured (discrete points) and calculated (curves) mass release as a function of residence time and oxygen concentration for bituminous (Kellingley) coal char particles using the L-H model with BO at (top) 1233 K and (bottom) 1673 K.

where k_i-s are the rate constants for the elementary steps (desorption and adsorption), and it can follow an Arrhenius law comprising the reaction mechanism. The values for activation energies of adsorption and desorption reactants E_i and their pre-exponential factors A_i for different coal types can be found in the literature. Table 4.10 summarises some of these values for atmospheric and pressurised conditions.

Char-CO_2 reaction model validation

The char-CO_2 reaction models were validated against experimental results obtained by different authors for three different coal chars tested under conditions similar to an industrial pulverised coal flame. The coal and the resulting char properties are given in Tables 4.11 and 4.12.

The first test used Indonesian subbituminous coal, called Roto. Char was prepared in the reactor in a nitrogen environment. A pressurised drop tube furnace reactor has been used to determine the rate of char-CO_2 gasification in the operating condition of an entrained flow gasification process. The objective of this investigation was to determine the effect of gasification temperature, partial pressure of CO_2, and total

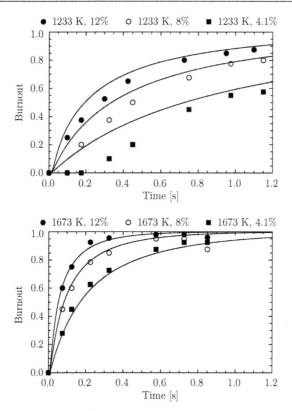

Figure 4.11 Char-O_2: Comparison of measured (discrete points) and calculated (curves) mass release as a function of residence time and oxygen concentration for bituminous (Kellingley) coal char particles using the first-order apparent kinetics model with BO at (top) 1233 K and (bottom) 1673 K.

system pressure on the rate of char-CO_2 gasification and to develop a suitable kinetic equation to predict the gasification rate. More details concerning the experiment are given by Ahn [194]. Since devolatilisation conditions greatly affect char characteristics, the chars were prepared at the same reactor conditions. The numerical results obtained for different reactor temperatures with an n-th-order kinetic model (given in Appendix A.1.1) are compared with experimental data in Figure 4.13.

There is good agreement between predictions and measurements for high-temperature (1473–1673 K) conditions. At lower temperatures (1273–1373 K), however, the predictions slightly overestimate the burnout rate. A large discrepancy is obtained for the lowest temperature investigated, namely, 1173 K.

Further, the effect of the system pressure on the gasification rates was investigated by varying the pressure to 5, 10, and 15 bar and keeping the reactor temperature constant at T = 1573 K. The predicted rate curves (Figure 4.14) follow the experiments, thus reproducing the decrease of the gasification rate with increase of the reactor pressure.

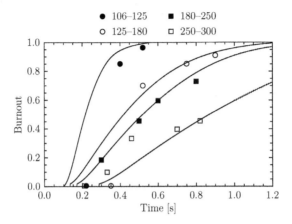

Figure 4.12 Char-O_2: Comparison of measured (discrete points) and calculated (curves) mass release as a function of residence time and particle size (in μm) for subbituminous coal char particles using the first-order apparent kinetics model with BO at 1173 K.

Table 4.9 The n-th-order model parameters for the char-CO_2 reaction.

Model	T [K]	k_0	E_A $10^6 \left[\frac{J}{kmol}\right]$	n [-]	Ψ [-]	Ref.
Halforder	–	$7,32 \cdot 10^{-2a}$	115	0,5	–	[195]
Firstorder	≤ 1223	$1,35 \cdot 10^{-4a}$	135,5			
	> 1223	$6,35 \cdot 10^{-3a}$	162,1	1	–	[196]
n-th - BA	–	$2,54 \cdot 10^{7b}$	257	0,56	26	[197]
n-th - DL	–	$1,12 \cdot 10^{8b}$	240	0,48	1	[197]
n-th - S	–	$1,23 \cdot 10^{9b}$	261	0,49	0,1	[198]
n-th - NL	≤ 1473	$1,03 \cdot 10^{9b}$	283	0,54	3	
	> 1473	$6,78 \cdot 10^{4b}$	163	0,73	3	[198]
n-th - BKK	–	$5,96 \cdot 10^{6c}$	257	0,3	–	[199]
n-th - Bio	–	$3,9 \cdot 10^{4d}$	215	0,38	–	[86]
n-th - Roto	–	$174,1^e$	71,5	0,4	–	[194]
n-th - Montana Rosebud	–		97,07	0,26	–	[200]

For k_0:

a $\left[\frac{kg}{m^2 \cdot s \cdot Pa^n}\right]$.

b $\left[\frac{1}{s \cdot MPa^n}\right]$.

c $\left[\frac{m^{3n}}{kg \cdot s \cdot kmol^{n-1}}\right]$.

d $\left[\frac{1}{s \cdot Pa^n}\right]$.

e $\left[\frac{1}{s \cdot MPa^{n+m}}\right]$.

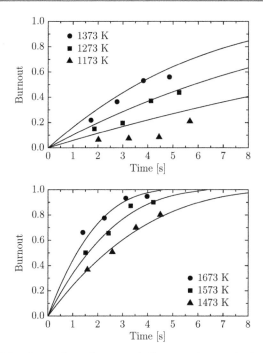

Figure 4.13 Char-CO_2: Comparison of measured (discrete points) and calculated (curves) mass release as a function of residence time for subbituminous (Roto) coal char particles at low (top) and higher (bottom) temperatures using n-th-order apparent kinetics model.

Finally, the concentration of CO_2 in the reactor was varied to 10, 20, 40, and 50 vol.% by keeping the reactor temperature and pressure constant (1573 K and 10 bar). The model results are in good agreement with the measurements for high-CO_2 partial pressure; however, they underpredict the reaction rates at lower concentrations, as shown in Figure 4.15.

The second test used char produced from a pulverised Montana Rosebud coal in a muffle tube furnace at 1365 K and 1 atm total pressure. The char was gasified in a drop tube reactor at 1473, 1773, and 2113 K gas temperatures. More details concerning this experiment can be found in Derschowitz [200]. The model used is an n-th-order kinetics model with parameters defined for Roto coal char that is of similar quality. The predictions obtained for 26 and 58 vol% CO_2 are in good agreement with the measurement data, as shown in Figure 4.16.

The third test considers gasification of coal char particles in an N_2-atmosphere with 40 vol% CO_2 at 1673 K and 5 bar reactor temperature and pressure, respectively. More details concerning these experiments can be found in Kajitani *et al.* [197]. Four different models have been used to predict the carbon conversion rate: first order [196], half order (here called CBK) [195], n-th, order, and L-H approximation [197]. Details concerning the model parameters are given in Appendix. A.1.1. The model predictions shown in Figure 4.17 are in good agreement with measurement data, with slightly over-predicted rates when using data for graphite (first-order model).

Table 4.10 Langmuir-Hinshelwood model parameters for the char-CO_2 reaction.

Coal	S_0 $\left[\frac{m^2}{g}\right]$	Ψ $[-]$	$k_{0,1}$ $\left[\frac{1}{s\cdot Pa}\right]$	$E_{A,1}$ $10^6\left[\frac{J}{kmol}\right]$	$k_{0,2}$ $\left[\frac{1}{Pa}\right]$	$E_{A,2}$ $10^6\left[\frac{J}{kmol}\right]$	$k_{0,3}$ $\left[\frac{1}{Pa}\right]$	$E_{A,3}$ $10^6\left[\frac{J}{kmol}\right]$	$k_{0,4}$ $\left[\frac{1}{s\cdot Pa^2}\right]$	$E_{A,4}$ $10^6\left[\frac{J}{kmol}\right]$	Ref.
BA	200	26	$27{,}54^a$	251	$4{,}93\cdot 10^{-7}$	-23	$5{,}72\cdot 10^{-7}$	$-48{,}1$	$2{,}15\cdot 10^{-16}$	$68{,}5$	[197]
DL	190	1	$1{,}28^a$	222	$3{,}17\cdot 10^{-6}$	-23	$1{,}273\cdot 10^{-6}$	$-48{,}1$	$2{,}15\cdot 10^{-16}$	$68{,}5$	[197]
Lig	–	1	$1{,}34\cdot 10^3$	$237{,}81$	$7{,}43\cdot 10^{-4}$	$29{,}52$	$2{,}96\cdot 10^{-15}$	$-235{,}3$	–	–	[87]
Pit	–	8	$116{,}457$	$257{,}49$	$2{,}16\cdot 10^{-5}$	$-2{,}512$	$6{,}03\cdot 10^{-3}$	$47{,}73$	–	–	[87]

$^a\left[\frac{g}{m^2\cdot s\cdot Pa}\right]$.

Table 4.11 Proximate and ultimate analysis of the investigated coals.

Seam Name Coal Rank	Roto[194] Subbituminous	Montana[200] Subbituminous	Chinese DL[197] Subbituminous
Proximate analysis (wt%, dry)			
Volatile	49.23	37.7	36.1
Fixed carbon	49.23	52.32	57.9
Ash	1.54	9.98	6.0
Ultimate analysis (wt% d.a.f.)			
Carbon	70.0	72.6	80.8
Hydrogen	5.2	4.8	4.8
Nitrogen	1.0	1.0	1.0
Oxygen	23.6	20.3	13.3
Sulphur (+ chlorine)	0.2	1.3	0.1

Table 4.12 Properties of coal char for three different coals.

Seam Name	Roto	Montana	Chinese DL
Density (kg/m^3)	990	992	990
Diameter (μm)	55	100	40
Carbon in char (%)	92.2	78.5	90.61
Temperature (K)	Varies	Varies	Varies
CO_2 conc. (vol%)	5	Varies	Varies

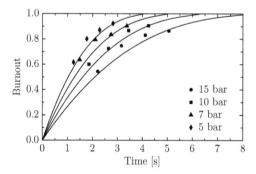

Figure 4.14 Char-CO_2: Comparison of measured (discrete points) and calculated (curves) mass release as a function of residence time and total pressure for subbituminous (Roto) coal char particles using n-th-order apparent kinetics model.

Char-steam reaction models

The reaction of carbon with steam, as in Eq. (3.6), has been shown to be similar to the reaction with CO_2 in terms of both relative rate and mechanisms. Some commonly used rate equation parameters for the steam gasification reaction are listed in Table 4.13.

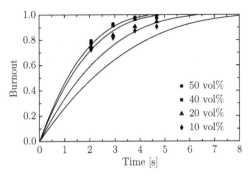

Figure 4.15 Char-CO_2: Comparison of measured (discrete points) and calculated (curves) mass release as a function of residence time and of CO_2 partial pressure for subbituminous (Roto) coal char particles using n-th-order apparent kinetics model.

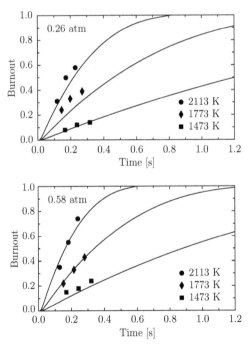

Figure 4.16 Char-CO_2: Comparison of measured (discrete points) and calculated (curves) mass release as a function of residence time and of temperatures at 0.26 atm (top) and 0.58 atm (bottom) CO_2 partial pressure for subbituminous (Montana Rosebud) coal char particles using an n-th-order apparent kinetics model.

Char-steam reaction model validation

The char-steam reaction models were validated with the experimental results obtained for Montana Rosebud coal chars against properties already given in Tables 4.11 and 4.12 under conditions similar to an industrial pulverised coal flame. Three different reactor

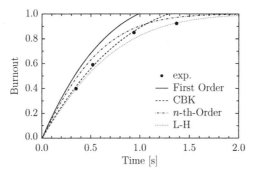

Figure 4.17 Char-CO_2: Comparison of measured (discrete points) and calculated (curves) mass release as a function of residence time for Chinese sub-bituminous (DL) coal char particles using n-th-order apparent kinetics and L-H models.

Table 4.13 nth-Order Model parameters for char-steam reaction.

Model	T [K]	k_0 $\left[\frac{kg}{m^2 \cdot s \cdot Pa^n}\right]$	E_A $10^6 \left[\frac{J}{kmol}\right]$	n [-]	Ψ [-]	Ref.
Half order	–	$7,82 \cdot 10^{-2}$	112,5	0,5	–	[195]
Firstorder	≤ 1233	$3,19 \cdot 10^{-1}$	208,016			
	Above 1233	$1,92 \cdot 10^{-3}$	146,992	1	–	[196]
nth - Montana Rosebud	4.14	4.14	109,62	1,19	–	[200]

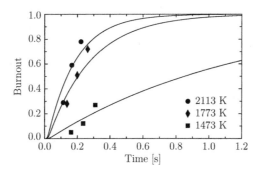

Figure 4.18 Char-H_2O: Comparison of measured (discrete points) and calculated (curves) mass release as a function of residence time and particle temperature for subbituminous coal char particles using first-order apparent kinetics model at 0.65 atm steam partial pressure.

temperatures—1473, 1773, and 2113 K—and two different steam concentrations—33.5 and 65 vol%–were considered. The numerical results obtained with an n-th-order apparent kinetic model are presented in Figures 4.18 and 4.19 for 65 vol% and 33.5 vol% steam concentrations, respectively.

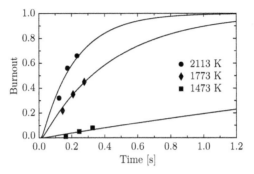

Figure 4.19 Char-H_2O: Comparison of predicted (curves) and measured (discrete points) mass release as a function of residence time and particle temperature for subbituminous coal char particles using first-order apparent kinetics model at 0.335 atm steam partial pressure.

Table 4.14 Pre-exponential factors for the Montana Rosebud nth-order model for 33.5 vol% and 65 vol% steam concentrations.

Temperature [K]	$k_{0,0.335}$ $\left[\frac{1}{s \cdot Pa^n}\right]$	$k_{0,0.65}$ $\left[\frac{1}{s \cdot Pa^n}\right]$
1473	$7,0 \cdot 10^{-3}$	$1,2 \cdot 10^{-2}$
1773	$1,6 \cdot 10^{-4}$	$1,2 \cdot 10^{-2}$
2113	$1,0 \cdot 10^{-4}$	$5,5 \cdot 10^{-3}$

The pre-exponential coefficients had to be adjusted to obtain good agreement with the measurements. These coefficients are listed in Table 4.14.

4.3   Modelling of thermal radiation

In calculating practical problems, the radiative heat transfer in the combustion system must be considered. Therefore, a radiative transport equation (RTE) must be added to the system of Eqs. 4.1–4.5. In terms of an Eulerian coordinate system, the RTE can be written as:

$$\frac{1}{c}\frac{\partial I_v}{\partial t} + \left(\nabla\Omega\right) I_v = k_v \left[4\pi \, I_{bv}(T) - I_v\right] + k_{v,p}\left[4\pi . I_{bv,p}(T) - I_{vp}\right] + S_{v,scat}$$

(4.78)

The left side of the equation represents the rate of change of spectral radiant energy density I_v in time and space. The first two terms on the right side of Eq. (4.78) represent the net rate of loss or gain of radiant energy from an element of matter per unit volume and unit frequency. The term $4\pi k_v I_{bv}$ represents the local rate of gas emission, and $k_v I_v$ represents the local rate of gas absorption of radiation per unit volume and unit frequency. When solid particles are to be considered in the system, similar terms appear for the particle emission and absorption, respectively. The last term in Eq. (4.78) $S_{v,scat} = k_s \left[\frac{1}{4\pi} \int \Phi_v \left(\Omega, \Omega' \right) I_v' \left(\Omega' \right) d\Omega' - I_v \right]$, describes the in-scattered radiation. The particle absorption coefficient for the entire spectrum can be expressed as:

$$k_p = \sum_{N}^{n=1} \epsilon_{p,n} \frac{A_{p,n}}{V} \tag{4.79}$$

where $\epsilon_{p,n}$ is the particle normal emissivity and $A_{p,n}$ is the particle surface area, m^2. Thus, the effect of gaseous species and particles on the absorption coefficient κ and on the scattering coefficient κ_s can be assumed as additive ($\kappa = \kappa_g + \kappa_p$ and $\kappa_s = \kappa_{s,g} + \kappa_{s,p}$, where subscripts "$g$" and "$p$" denote gas and particles, respectively).

The calculation of the spectral radiative heat transfer problem as part of the overall CFD-based model is computationally far too expensive, and therefore the RTE is integrated over the wavelength.

To solve the RTE, several methods have been introduced:

- The zone method, proposed by Hottel and Sarofin [3], divides the system into two types of zones with homogeneous distribution of the parameters (volume and surface) and calculates the exchange of the radiative energy between them.
- Statistical or Monte Carlo types of methods use a finite number of rays to represent the radiative energy.
- Flux methods can be divided into multiflux methods (two-, four-, or six -flux approach) and the discrete ordinates (DO) method (radiation at each control volume is divided into discrete directions and the intensity is calculated along the CFD grid).
- Discrete transfer method (DTM), proposed by Lockwood and Shah [201], calculates the radiation intensity along selected ray directions (a combination of Monte Carlo, zone method, and DO method).

Although Hottel's Zone method and the Monte Carlo technique have been accepted as the most accurate methods for calculating radiative transfer, these traditional methods have not found application in comprehensive combustion modelling due to their large computational time and storage requirements. The flux methods are preferred for combustion modelling because the simplified radiation transport equation can be simultaneously solved with the differential equations of the fluid flow.

4.3.1 Radiative properties of walls

The zero transmittance of the furnace wall and the application of Kirchhoff's law yield the following relation between reflectance ρ_w and emittance ϵ_w at the wall:

$$\rho_w + \epsilon_w = 1 \tag{4.80}$$

Determining the wall emissivity is a complicated task, since the rough furnace walls are generally covered to a varying degree with deposits, such as soot and ash. The radiative properties of those deposits are dependent on their chemical composition and can vary not only as function of temperature, location, and wavelength, but also as a function of furnace operating time. Normally, emissivity ϵ_w at furnace walls can be assumed to be in the range of 0.6–0.8. The surface emission and reflection are commonly assumed to be diffuse, which is a reasonable approximation for rough boiler walls. For the metal parts of the burner, quarl wall, and the like, typical values for ϵ_w are 0.2–0.3 and are treated as diffuse as well.

4.3.2   Modelling of gas emissivity

Combustion gases do not scatter radiation significantly, but they are strong, selective absorbers and emitters of radiant energy. Therefore, spectral variation of the radiative properties must be accounted for.

Among several models suggested, the exponential wideband model (EWBM) [202] was found to be the most accurate over the wide range of partial pressures and temperatures that are important in combustion systems. In the past, this model was considered quite complicated and too detailed for describing the radiative heat transfer in furnaces. Therefore, simpler models that account for the typical gas partial pressures and temperatures for air combustion systems, such as the weighted sum-of-grey-gases, assuming three-grey-one-clear gas [187] and some polynomial expressions [203] are commonly used in literature.

According to the weighted sum-of-grey-gases model, first proposed by Hottel and Cohen [204], the total emissivity ϵ_g of a gas layer with thickness L and at temperature T can be defined as:

$$\epsilon_g = \sum_{i=0}^{G} a_{\epsilon,i} \left[1 - e^{-\kappa_i p L} \right] \tag{4.81}$$

where G is the number of grey gases. The weighting factors $a_{\epsilon,i}$ may be interpreted as the fractional amounts of blackbody energy in the spectral regions where the grey gas absorption coefficient $\kappa_{g,i}$ exists, and they are functions of temperature:

$$\sum_{i=0}^{G} a_{\epsilon,i} = 1, \quad a_{\epsilon,i} = b_{i,0} + \sum_{j=1}^{J} b_{i,j} T^{j-1} \tag{4.82}$$

where $b_{i,j}$ are coefficients for the J-th order polynomial in temperature for emissivity.

Usually the absorption coefficient for $i = 0$ is assigned a value of zero to account for the transparent windows in the spectrum. The number of spectral regions of the weighted-sum-of-grey-gases model usually ranges from two to four. Eq. (4.78) can be solved over the whole integration domain once per time step (or iteration) for each spectral region. Smith et al. [187] presented polynomial coefficients and absorption coefficients for different ratios of partial pressure of water vapour p_{H_2O} to partial pressure of carbon dioxide p_{CO_2} for $p_{H_2O}/p_{CO_2} = 1 \& 2$. The model of Smith et al. is valid within the temperature range 600–2400 K and within the partial pressure × path length range $0.001 - 10$ atm × m.

The gas absorption coefficient k_g is calculated over the whole spectrum, as given in the following equation:

$$\kappa = -\frac{1}{L} ln \left(1 - \epsilon\right) \tag{4.83}$$

The path length, L, is approximated using a characteristic size of the computational cell defined as $L = V_{cell}^{1/3}$ with V_{cell} being the volume of computational cell. In oxy-combustion, the gas emissivity will differ from air-firing systems primarily due to different partial pressures of H_2O and CO_2 but also due to different ratio of partial pressures of the two gases. Hence the use of the original weighted sum-of-grey-gases model when modelling oxyflames can lead to deviations from reality. Therefore, the model coefficients were modified by Gupta et al. [121], Johansson [120], and Payne et al. [122] in order to account for the changed partial pressures.

A different approach was followed by Erfurth et al. [124], introducing a nongrey implementation of the EWBM for modelling oxy-firing systems as follows:

Since combustion gases are highly nongrey and significant gradients in gas and wall temperatures are sure to be incurred, Erfurth et al. [124] suggested modelling the radiation transport in a nongrey fashion, that is, by solving the RTE for several wavenumber intervals Δv_i. To model the local absorption coefficients κ_i of the gas mixture, the greyband version of the EWBM by Edwards and Menard [202] was used. The model calculates constant transmissivities τ_m and an equivalent bandwidth Δv_m based on local temperature T_{gas}, partial pressure p_i of species i associated with the band m, total pressure P_t, and a travelled path-length L.

The transmissivity is calculated for individual bands of the species CO_2, H_2O, and CO, as listed in Table 4.15, along with their spectral locations. Some of these bands are located close to each other, which leads to overlap effects. These are taken into account

Table 4.15 Radiation bands i and intervals m chosen to solve separate RTEs when using EWBM.

Species j	Band i	Location [1/cm]	Intervals				
			m	$v_{l,m}$	$v_{u,m}$	$\lambda_{l,m}$	$\lambda_{u,m}$
CO_2	1	960	1	3219	4301	2.3251	3.107
	2	1060	2	4301	5142	1.945	2.325
	3	2410	3	5143	5528	1.809	1.994
	4	3660	4	5529	7105	1.407	1.809
	5	5200	5	7106	7394	1.3521	1.407
H_2O	6	1600	6	890	1115	8.969	11.236
	7	3760	7	1116	1871	5.345	8.961
	8	5350	8	1872	2410	4.149	5.342
	9	7250	9	2411	3218	3.108	4.148
CO	10	2143	10	7395	33333	0.3	1.352
	11	4260					

by calculating total transmissivities $\tau_{t,l}$ from the band transmissivities $\tau_{m,l}$:

$$\tau_{t,l} = \sum \tau_{m,l} \qquad (4.84)$$

Here l is a spectral region whose boundaries are marked by either the end or the beginning of one of the bands m. $\tau_{m,l}$ are the transmissivities of those bands. $\tau_{t,l}$ is thus the total transmissivity of the region l.

The spectrum was divided into 10 intervals Δv_i. The 11 bands listed in Table 4.15 are concentrated in six spectral regions, which were thus selected as wavenumber intervals. The gaps between those regions, where combustion gases are transparent, were chosen as the remaining four intervals to take into account particle radiation in those regions. No RTE was solved for low wavenumbers, since for the temperatures of gases and particles only a negligible fraction of black-body energy is emitted there. Consequently, the H_2O and CO_2 bands at low wavenumbers are not listed in Table 4.15. The same is true for very high wavenumbers. The 10 intervals were chosen to cover the spectrum from 900 to 33333 cm^{-1}.

The overall transmissivity τ_i in Δv_i can be calculated as:

$$\tau_i = \frac{1}{\Delta v_i} \int_{\Delta v_i} \tau_t dv \qquad (4.85)$$

From τ_i the absorption coefficient κ_i is readily calculated using the Beer-Lambert law:

$$\kappa_i = \frac{1}{L} ln \frac{1}{\tau_i} \qquad (4.86)$$

The proposed nongrey EWBM model was validated against narrowband model predictions in respect to its implementation and integration into the discrete ordinates method [124]. Full validation, however, can be committed when large-scale oxy-firing experimental data become available.

4.4   Summary of Chapter 4

Engineering calculations and numerical simulations of coal combustion processes are the "connecting part" between combustion theory and practice. Such predictions require an accurate rate expression for reactions. Therefore, a CFD-based numerical model, including improved combustion models for oxyfuel flames, was developed and presented.

Turbulence modelling remains the main issue to be solved in order to perform correct predictions of swirl flames. Although swirling flows and their impact on mixing in the near-burner zone and thus on the flame stability are important in all combustors, they play a more significant role in the case of oxycombustion. The performance of the standard economical model approaches such as RANS models was addressed in Section 6.4.1, where a cold flow in an oxycoal swirl burner was simulated and validated against experiments.

Due to its simplicity and robustness the eddy dissipation concept model for turbulence combustion seems to be the most practical choice for modelling industrial-scale pulverised coal flames. The model should be combined with a chemical kinetics mechanism for the gas phase reactions that take place in an oxycombustion and that capture the chemical effects of CO_2.

A new algorithm for coal particle modelling was introduced. Due to the higher partial pressures of oxygen, carbon dioxide, and water vapour in oxyfuel than in air combustion, the proposed model is based on the assumption that the devolatilisation and the char oxidation/gasification run in parallel, thus overlaping in time.

Empirical models for coal particle devolatilisation were tested against experimental data showing correct correlation of the rate of particle weight loss with empirical kinetics and estimated ultimate volatile yield. However, they all lack in generality of coal types and reaction conditions and cannot give detailed information about devolatilisation for gas phase flame reactions because the models were established on limited reaction mechanisms. Therefore, advanced structural models were examined, and the CPD model was tested with experimental data, showing good predicting capabilities without introducing any parameter fitting.

State-of-the-art mechanistic models for char burnout were analysed, and for the three types of kinetics considered, that is, the apparent kinetics, the intrinsic kinetics, and Langmuir kinetics, only one kinetic parameter was adjusted to enable the representation of the char burnout process by varying the temperature, partial pressures, particle diameter, and coal quality. For oxyfuel flame conditions, three main heterogeneous reactions were identified and empirical kinetics for these reactions were validated for char-O_2, char-CO_2, and char-H_2O.

Gas emissivity modelling should consider the changed composition of the flue gas during oxyfuel coal combustion. The exponential wideband model seems to be the natural choice for oxyfuel applications due to its accuracy over a wide range of partial pressures and temperatures typical for oxycoal flames.

The improved coal particle models were integrated in a CFD-based numerical model for pulverised coal combustion. Thus the model was used for oxyflame simulations and actively applied in the process of design and development of swirl oxyflame burners, as shown in Chapter 6.

Part II

Experiments

5 Gaseous Combustion in CO_2/O_2 Atmosphere

This chapter compiles the experimental and numerical combustion investigations of gaseous fuel in N_2/O_2 and CO_2/O_2 atmospheres using a 25 kW gas reactor. The specific objective here is to assess the importance of the chemical effects of the high CO_2 concentration in oxyfuel combustion of methane under well-defined reaction conditions in flameless combustion. The advantage of using flameless combustion lies in the possibility to "isolate" the rest of the effects (e.g., molar heat capacity, CO_2 dissociation or thermal radiation) on the combustion rate by keeping the reactor temperature constant through control by a cooling system. Furthermore, stable combustion of methane with less than 21 vol% oxygen in the CO_2/O_2 mixture can be achieved, as reported by Heil [205].

Flameless combustion burners are well known in industrial high-temperature combustion applications [206–208]. Due to the high momentum of the air inflow, intense mixing of fuel, oxidiser, and hot flue gas occurs inside the combustion chamber. As a result, there is no flame attached to the burner. Instead, a large reaction zone with low oxygen content is generated, thus promoting "distributed" reactions in the combustion chamber. The temperature and species distribution in the combustion chamber are uniform so that this combustion mode is considered to behave similarly as a perfectly stirred reactor [209,210].

5.1 Test facility for flameless gas combustion

5.1.1 Experimental setup

Figure 5.1 shows a sketch of the test facility at the Institute of Heat and Mass Transfer at RWTH Aachen University; this facility is used for flameless gas combustion experiments. The furnace has a vertical, cylindrical geometry with a height of 1 m and an inner diameter of 0.55 m. The combustion chamber is air-cooled to obtain a constant inner temperature without reducing the thermal load. This is achieved by cooling pipes that are mounted close to the wall.

The burner used is a commercial recuperative burner designed for flameless operation at a thermal load of 25 kW. It is installed at the bottom of the combustion chamber facing upward to minimise the influence of gravity on the flow. The hot flue gas leaves the combustion chamber through the recuperator of the burner, thus heating the incoming fresh gas mixture. Methane enters the furnace through a central opening in the

Combustion of Pulverised Coal in a Mixture of Oxygen and Recycled Flue Gas. http://dx.doi.org/10.1016/B978-0-08-099998-2.00005-9

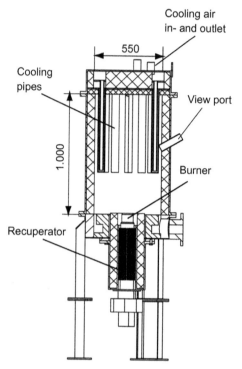

Figure 5.1 Scheme of the test facility for flameless gas combustion, 25 kW.

front plate of the burner. This central opening is surrounded by six openings that supply the preheated oxidiser gas mixture (see Figure 5.2). The burner and the experimental burning chamber were designed by WS Wärmeprozesstechnik GmbH.

5.1.2   Oxyfuel tests

A series of tests with different oxidiser mixtures has been carried out at a constant furnace temperature of 900 °C. Four different gas mixtures consisting of N_2/O_2 and CO_2/O_2, each with 21 vol% O_2 and 18 vol% O_2 (Air-21, Air-18, Oxyf-21, Oxyf-18), were used as oxidiser streams, as shown in Table 5.1. All experiments were conducted at an oxygen/fuel ratio of 1.15, resulting in a constant volume flow of oxygen (5.78 m_N^3/h) for all tests.

To investigate the influence of the various gas mixtures on the species and temperature distribution in the combustion chamber, detailed in-flame measurements were conducted through an opening in the top of the combustion chamber. With a traverse system a probe was moved in an axial range of 0.05 m to 0.95 m from the burner outlets and a radial range from 0 m to 0.15 m from the center of the furnace. In order to determine the influence of the composition of the oxidiser on the species distribution, an air-cooled probe was used to take gas samples at different positions in the combustion chamber. The concentrations of O_2, CO_2, and CO at 10 radial and 4 axial positions

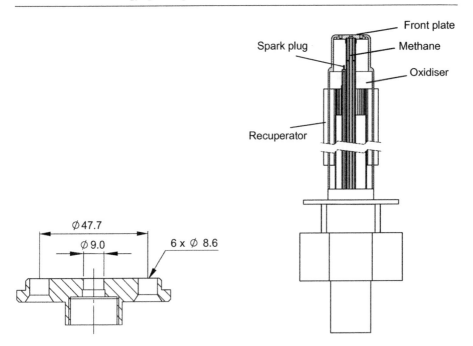

Figure 5.2 Scheme of the burner for flameless gas combustion.

Table 5.1 Oxidiser composition (O_2 volume flow was 5.78 m_N^3/h for all cases).

	N_2 Volume Fow [m_N^3/h]	CO_2 Volume Fow [m_N^3/h]
Air-18	26.35	0
Air-21	21.76	0
Oxyf-18	0	26.35
Oxyf-21	0	21.76

Source: Reprinted from Proceedings of the Combustion Institute, Vol.33/2, P. Heil, D. Toporov, M. Förster, R. Kneer, Experimental investigation on the effect of O_2 and CO_2 on burning rates during oxyfuel combustion of methane, 3407–3413, (2011), with permission from Elsevier.

(a total of 40 points) were measured. Additionally, another traversable probe with a thermocouple was used to measure the gas temperature at the same positions. CO and CO_2 were measured with nondispersive ultraviolet absorption; the measurement principle of the O_2 analyser is of the paramagnetic type. For temperature measurements, a PtRh/Pt thermocouple type S was used. More details about the measurement equipment and the data evaluation are given by Heil [205].

To minimise the influence of the measurement's sequence, the order of the measurement position was selected randomly. The presented radial profiles were obtained during two measurements performed on two different days using two different sequences of measurement positions. For each position, the measurement time selected was two minutes. The measured data were evaluated by variance analysis. To improve clarity, the confidence intervals are not shown in the graphs. They are presented in the next section.

5.2 Experimental results

All the experiments were performed in flameless oxidation mode, which was confirmed by optical observation. The experimental results show stable combustion and a pronounced external recirculation zone, thus corresponding to the definition of flameless oxidation given by Wünning [211].

The globally measured mean values of oxygen at the combustion chamber exit are 2.7 vol% O_2 for Air-18 and Oxyf-18 and 3.0 vol% O_2 for Air-21 and Oxyf-21. The measured global CO levels were below the detection limit for all cases, showing full burnout.

The measured values of flue gas temperature and gas concentrations, averaged over 40 measurement points inside the combustion chamber, are given in Table 5.2. The corresponding standard errors were $\pm 5.6\,°C$, ± 0.08 vol%, and ± 31 ppm for temperature, O_2, and CO concentrations, respectively.

The relatively small differences between the averaged temperature levels indicate constant temperature conditions during all experiments. The differences in the temperature levels between N_2/O_2 and CO_2/O_2 cases can be attributed to (1) precision of the experiment and of the measurement device, and (2) the higher emissivity/absorptivity of CO_2 compared to N_2. However, the short path lengths within the furnace should minimise the effect of the changed gas emissivity.

The difference in the O_2 concentrations in the oxidiser predetermines the different average O_2 levels obtained inside the combustion chamber. However, the difference of 3 vol% O_2 in the oxidiser causes a difference of about 1.2 vol% for air combustion and only about 0.4 vol% for oxyfuel combustion. Concerning the CO levels, it can be seen that for air combustion there is only a small difference between the cases of Air-21 and Air-18. In oxyfuel combustion, however, the CO concentration is about twice as much

Table 5.2 Measured values of flue gas temperature and gas concentrations averaged over 40 measurement points inside the combustion chamber.

	Temperature [°C]	O_2 [Vol%], Dry	CO [ppm], Dry
Air-18	903	4.2	196
Air-21	877	5.4	206
Oxyf-18	926	4.4	1359
Oxyf-21	904	4.8	591

Source: Reprinted from Proceedings of the Combustion Institute, Vol.33/2, P. Heil, D. Toporov, M. Förster, R. Kneer, Experimental investigation on the effect of O_2 and CO_2 on burning rates during oxyfuel combustion of methane, 3407–3413, (2011), with permission from Elsevier.

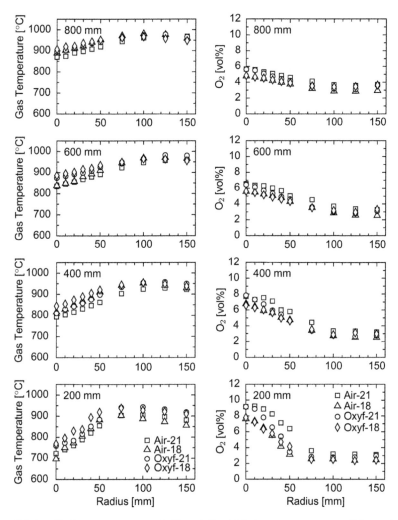

Figure 5.3 Detailed gas temperature (left) and O_2 (right) measurements at different burner distances.
Source: Reprinted from Proceedings of the Combustion Institute, Vol.33/2, P. Heil, D. Toporov, M. Förster, R. Kneer, Experimental investigation on the effect of O_2 and CO_2 on burning rates during oxyfuel combustion of methane, 3407–3413, (2011), with permission from Elsevier.

for Oxyf-21 and about six times higher for Oxyf-18 as those obtained for air combustion. The interpretation of these results requires a more detailed look at the measured data.

The measured radial temperature profiles are plotted in Figure 5.3. The 95% confidence interval for all measurement points is $\pm 20\,°C$. The radial profiles of the temperature measured for all cases are very similar, thus showing similar heat release due to combustion. Close to the burner, at an axial distance of 200 mm, the temperature difference between the radial positions 0 mm and 150 mm is about 200 °C, reflecting the initial heating of the incoming fuel-oxidiser mixture. This difference decreases with increased distance from the burner, approaching 50 °C at 800 mm axial distance. The slow rise of gas temperature and the small temperature gradients suggest a continuous reaction, which is typical for flameless combustion. Between a radial position of 100 mm and the edge of the combustion chamber, the flue gas is recirculated back to the burner. This can also be seen from the O_2 profiles shown in Figure 5.3 (95% confidence interval ± 0.05 vol%).

For all settings and axial positions, the maximum of the O_2 concentration is located in the middle of the combustion chamber, showing the main fuel/oxidiser jet. Close to the burner, at an axial distance of 200 mm, the measurements show a decrease of O_2 concentrations in the region between the center line and 100 mm radial distance. In the outer region, between 100 mm and 150 mm, the O_2 profile becomes uniform, thus showing the recirculation zone. With increased distance from the burner, the O_2 profiles flatten.

The measurements of gas temperature and O_2 concentrations show a reaction zone in the center of the combustion chamber between radial positions of 0 mm and 100 mm. In this area the gas temperature increases and the O_2 concentration decreases with increasing burner distance. The flue gas is recirculated back to the burner in the recirculation zone formed at radial positions lager than 100 mm. In this area the O_2 concentration is more uniform, indicating much slower reaction rates.

The CO concentrations measured for air and oxycombustion are given in Figure 5.4. The 95% confidence interval for all measurement points is ± 394 ppm. Higher CO concentrations are measured in the recirculation zone, where the CO oxidation actually takes place for both types of combustion (air and oxyfuel). However, the levels obtained for air and oxyfuel cases are quite different.

In the case of air combustion, the measured CO profiles overlap, thus demonstrating no influence of O_2 concentration on the CO production and oxidation rates when the temperature is constant. At a burner distance of 200 mm, the CO concentrations are negligibly low. This means that the production of CO in the main jet stream did not start yet and the oxidation of the already produced CO is completed in the recirculation zone. With increasing burner distance, the CO concentration increases, with the highest level (800 ppm) measured at an axial distance of 800 mm and a radial position of 150 mm.

In case of oxyfuel combustion, however, the measured CO profiles differ for the two investigated cases. Similar to air combustion, the measured CO levels for both settings at the axial position 200 mm, are low. At a burner distance of 400 mm, there is already a significant increase in CO levels for the Oxyf-18 case. The minimum values measured at this axial distance (about 200 ppm) are obtained in the center of the combustion chamber, whereas the maximum (about 3000 ppm) is obtained at a radial position of 150 mm. In comparison, the CO concentrations measured at this axial position for the Oxyf-21 case stay below 300 ppm. With increased burner distance, the level of CO increases for both settings, with CO levels for Oxyf-21 being about two times lower than those obtained for Oxyf-18. The higher CO levels obtained for Oxyf-18 in the reaction zone in the middle of the combustion chamber show higher CO production rates than in Oxyf-21. Furthermore, the CO profiles for Oxyf-18 measured in the recirculation zone between the burner distances 400 mm and 600 mm show almost identical values, whereas in the case of Oxyf-21, the CO levels drop rapidly. This shows a significant effect of the O_2 concentration on the oxidation rates of CO in oxyfuel combustion.

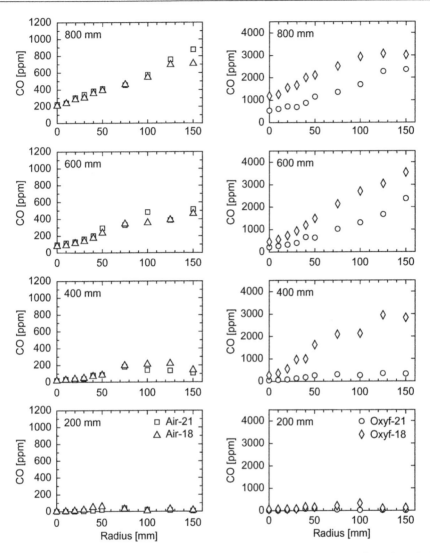

Figure 5.4 Detailed CO measurements at different burner distances for air (left) and oxyfuel (right) combustion.
Source: Reprinted from Proceedings of the Combustion Institute, Vol.33/2, P. Heil, D. Toporov, M. Förster, R. Kneer, Experimental investigation on the effect of O_2 and CO_2 on burning rates during oxyfuel combustion of methane, 3407–3413, (2011), with permission from Elsevier.

5.3 Numerical results

To assess the possibilities of mathematical models to predict the changes in combustion physics during oxyfuel, numerical simulations of the Oxyf-18 case were performed. The eddy dissipation model and the eddy dissipation concept, both combined with a

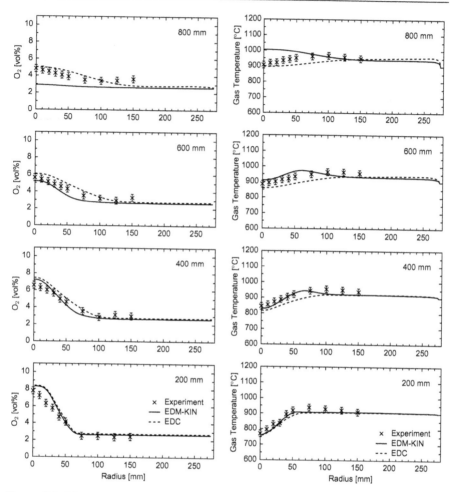

Figure 5.5 Measured and predicted oxygen concentration (left) and gas temperature (right) profiles at different axial burner distances for the Oxyf-18 case.
Source: Reprinted from Proceedings of the Combustion Institute, Vol.33/2, P. Heil, D. Toporov, M. Förster, R. Kneer, Experimental investigation on the effect of O_2 and CO_2 on burning rates during oxyfuel combustion of methane, 3407–3413, (2011), with permission from Elsevier.

simple kinetics mechanism, were used for modeling turbulence/chemistry interaction, as described in Section 4.1.2. The predicted gas temperatures, O_2 and CO concentrations at distances of 200 mm, 400 mm, 600 mm, and 800 mm from the burner outlets, were compared with experimental data.

The data obtained for gas temperature are shown on the right in Figure 5.5. The predictions of both models are quite identical, with the measured profile close to the burner (200 mm), where an intensive mixing between the incoming fresh gas mixture with recirculated hot flue gas takes place. Further downstream, however, at 400 mm, 600 mm, and 800 mm distance from the burner outlets, the EDM/kinetic model predicts

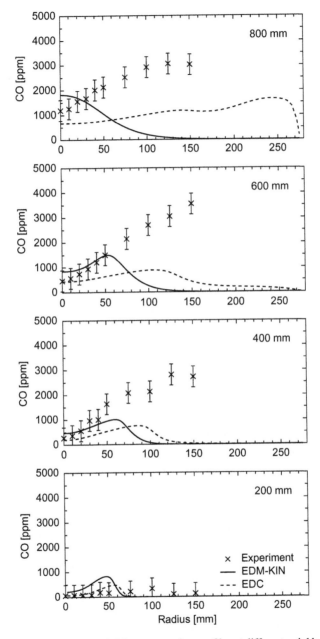

Figure 5.6 Measured and simulated CO concentration profiles at different axial burner distances for the Oxyf-18 case.
Source: Reprinted from Proceedings of the Combustion Institute, Vol.33/2, P. Heil, D. Toporov, M. Förster, R. Kneer, Experimental investigation on the effect of O_2 and CO_2 on burning rates during oxyfuel combustion of methane, 3407–3413, (2011), with permission from Elsevier.

higher temperatures near the axis. The O_2 profiles, on the left in Figure 5.5, show that in this region the main oxidation process takes place. Such temperature peaks are characteristic for flame like structures and differ from the measured values. In contrast, the predictions by the EDC correspond quite well to the experiments.

A comparison between the model predictions and experiments for the CO concentration is presented in Figure 5.6. As shown, the global two-step mechanism for methane oxidation, which was developed and successfully used for air combustion, fails in prediction of the intermediate species CO in the post-combustion zone in case of oxyfuel combustion [205]. This requires a new kinetic mechanism that should consider the chemical effects of CO_2, as discussed in detail in Section 3.4.

5.4   Summary of Chapter 5

Combustion of gaseous fuel in a CO_2 atmosphere differs from that in air. To define and clarify the exact mechanisms leading to this difference, an experimental investigation of flameless methane combustion in different atmospheres (N_2/O_2 and CO_2/O_2) and O_2 concentrations (21 vol% and 18 vol%) involving detailed in-flame measurements was carried out.

The measurements of gas temperature obtained for all settings show uniform temperature distributions and negligible differences between the cases. The measured O_2 concentrations show a similar trend in the radial profiles for all cases. The levels of the O_2 profiles differ depending on the O_2 concentration of the oxidiser.

The CO levels measured for N_2/O_2 atmosphere are identical, thus showing that there is no influence of O_2 concentration on the CO production and oxidation rates when the temperature is constant. In CO_2/O_2 atmosphere, however, an increase of O_2 concentrations from 18 vol% to 21 vol% leads to a reduction of almost two times the measured CO levels in the reaction zone, thus showing lower CO production rates than in Oxyf-18. Furthermore, the CO levels in the recirculation zone show that there is a significant effect of the O_2 concentration on the oxidation rates of CO in oxyfuel combustion.

Summarising, the results obtained and presented here demonstrate the following: By elimination of the influence of (1) molar heat capacity (by keeping furnace temperature constant for all cases), (2) CO_2 dissociation (by keeping furnace temperature at max 900 °C), and (3) thermal radiation (by performing experiments in a small-scale furnace at constant and low temperature), it can be estimated that the chemical effects of CO_2 presence have a significant impact on the production and consumption rates of carbon monoxide in oxyfuel combustion. An increase of O_2 in oxyfuel combustion can reduce this impact; however, further investigations, including detailed mathematical descriptions of this phenomenon, are needed to clarify and assess this effect.

Numerical predictions based on existing global kinetic mechanisms, coupled with a state-of-the-art turbulence combustion model, can provide an adequate simulation of methane oxidation in O_2/CO_2 atmosphere. However, new detailed mechanisms, considering the chemical effects of CO_2 on the CO-reaction rates, are needed.

6 Coal Combustion in CO_2/O_2 Atmosphere

This chapter compiles the experimental and numerical investigations of combustion of pulverised coal in both air and oxyfuel atmospheres that were performed at 100 kW pilot plant. The specific objectives here are focused on the realisation of a stable swirl oxycoal flame and the conduction of detailed in-flame measurements to obtain valuable information about the flame's characteristics and its behaviour.

The achievement of these goals goes through a systematic approach to applied science based on fundamental theory of coal combustion and fluid dynamics of swirl burners and considering the additional effects due to the high concentrations of carbon dioxide in oxycoal combustion. This approach is an integral part of this study and is described in this chapter.

6.1 Oxycoal pilot-scale furnace

The PF combustion experiments were performed in the oxycoal pilot plant of the Institute of Heat and Mass Transfer (WSA) at RWTH Aachen University, Germany schematically shown in Figure 6.1. The test rig is composed of (1) a vertical, cylindrical furnace with an adiabatic combustion chamber and, flue gas recirculation line; (2) flue gas cleaning device (ceramic filters) for particle removal; (4) heat exchanger; (5) flue gas recirculation fan, and (6) flue gas reservoir.

The furnace is a top-down firing with a length of the combustion chamber of 2.1 m and an inner diameter of 0.4 m. The burner is axially movable; thus detailed information of the flame properties can be obtained by optical and in-flame measurements at different distances from the burner orifice, as shown in Figure 6.2. Four ports for optical and probe measurements are located in the measurement plane. The ports are equipped with a gate valve system. Thus, probes can be changed and the windows can be cleaned without interrupting operation.

The burner is a swirl burner with single annular orifice (SAO) or single central orifice (SCO) arrangements through which the primary stream and pulverised coal are supplied. Because the primary stream is used as carrier fluid for the coal, its oxygen content was kept constant at 19 vol%. The secondary stream can be swirled, whereas the amount of swirl can be adjusted. It is injected into the combustion chamber through an annulus surrounding the primary stream inlet. A tertiary stream can be injected through an annulus, enclosing the quarl. Another gas stream, the staging fluid, enters the furnace through an annulus at the outer diameter of the furnace. Its main purpose

Combustion of Pulverised Coal in a Mixture of Oxygen and Recycled Flue Gas. http://dx.doi.org/10.1016/B978-0-08-099998-2.00006-0

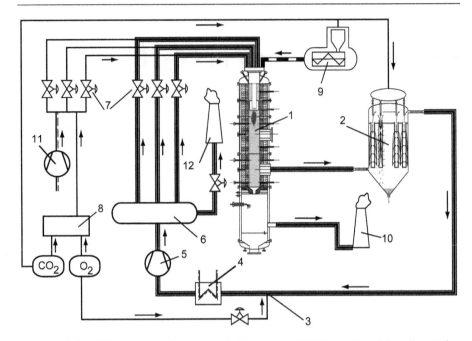

Figure 6.1 Simplified scheme of the oxycoal pilot plant at RWTH Aachen University: (1) furnace; (2) hot FG filtration unit; (3) oxygen addition; (4) FG cooler; (5) FG fan; (6) FG reservoir; (7) control valves; (8) gas mixing unit; (9) coal feeder; (10) flue stack; (11) air fan; (12) startup stack.

is to reduce the amount of gas, which is injected through the other inlet openings, and thus reduces axial velocities and the local stoichiometry at the burner.

The coal used for tests was pre-dried Rhenish lignite (see Tables 6.1 and 6.2).

Experiments with recycled flue gas were performed by opening the recirculation line. For this, a stable air flame was set up with volume flow rates through the burner selected as required for the oxycoal flame. Part of the flue gas (around 80% of the volume) was redirected from the combustion chamber and passed through the FG cleaning system to the FG reservoir by switching on the FG fan. Then the FG was released through the startup stack. Thus, the recirculation line was heated up to operation temperature (800 °C). When the operation temperature was reached, the flue gas was enriched with oxygen up to the required level. The switching from air to O_2 with recycled flue gas (O_2/RFG) was carried out by gradually replacing the air streams through the burner with the O_2/RFG mixture from the FG reservoir in steps of about 5% of the total volume flow. Simultaneously, the flow of RFG from the FG reservoir to the startup stack was reduced by the same step width of 5%, thus keeping constant the RFG flow rate through the recirculation line. This procedure was continued until reaching full replacement of the combustion air with RFG/O_2. During the switching procedure, the flame remained stable. The whole switching from air to oxycoal operation lasted about 15 min.

Thus, the facility can be operated in three different combustion modes, namely (1) air combustion; (2) combustion in CO_2/O_2 mixture (representing artificial dry recycle), and (3) combustion in O_2/RFG mixture with hot FG recirculation (representing wet recycling).

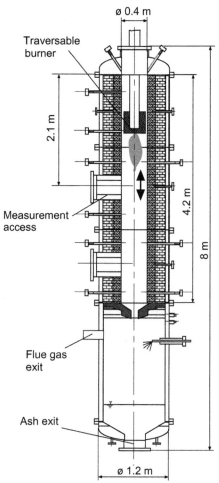

Figure 6.2 Scheme of the oxyfuel test furnace at the institute of heat and mass transfer at RWTH Aachen University.

The plant was operated in a shift regime, three persons per shift (one operating the plant, the second performing the measurements, and the third doing system controlling). Two days are needed for plant heatup and cooldown. Measuring of one variable at all planes takes around one or two days. Thus one complete measurement program could be done in a minimum of 10 days' nonstop failure-free run of the facility.

6.2 Burner design and flame stability

Several swirl burners were tested, with the geometries presented in Figure 6.3. The central orifice is used to supply the primary stream and coal. The secondary stream can be swirled, whereas the amount of swirl can be adjusted. Normally the secondary

Table 6.1 Proximate and ultimate analysis of rhenish lignite in mass -%.

	As Received	Dry	d.a.f.
Water	8.40	—	—
Ash	4.10	4.48	—
Volatiles	46.60	50.87	53.26
Char	40.90	44.65	46.74
Water	8.40	—	—
Ash	4.10	4.48	—
Carbon	67.40	73.58	77.03
Hydrogen	4.24	4.63	4.85
Oxygen	14.70	16.05	16.80
Nitrogen	0.86	0.94	0.98
Sulfur	0.30	0.33	0.34

Table 6.2 Particle size distribution. Upper class limit and percentage passing through.

d [μm]	0.90	1.10	1.30	1.50	1.80	2.20	2.60	3.10
Vol%	1.23	1.75	2.21	2.63	3.18	3.84	4.43	5.14
d [μm]	3.70	4.30	5.00	6.00	7.50	9.00	10.50	12.50
Vol%	6.00	6.89	8.00	9.74	12.66	15.92	19.44	24.23
d [μm]	15.00	18.00	21.00	25.00	30.00	36.00	43.00	51.00
Vol%	30.08	36.70	42.75	49.87	57.40	64.76	71.56	77.76
d [μm]	61.00	73.00	87.00	103.00	123.00			
Vol%	84.04	90.13	95.36	98.80	100.00			

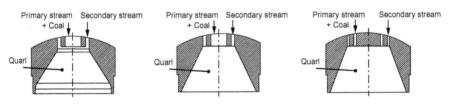

Figure 6.3 Burner A (left), burner Oxy-1 (middle), burner Oxy-2 (right).
Source: Reprinted from Fuel, Vol.88/7, P. Heil, D. Toporov, H. Stadler, S. Tschunko, M. Förster, R. Kneer, Development of an oxycoal swirl burner operating at low O_2 concentrations, 1269–1274, (2009), with permission from Elsevier.

stream was swirled, with a ratio[1] between the tangential and axial components of velocity $s = V_{tang}/V_{axial}$ being adjusted to 1.0. This ratio was obtained by numerical simulation of the swirl chamber of the burner. It is injected to the combustion chamber

---

[1]Usually the swirl level is given by the swirl number S, defined as the ratio between the axial flux of angular momentum G_ϕ and the axial flux of axial momentum G_x; $S = G_\phi/(G_x r_b)$, with r_b being the burner radius. Due to the specifics of the swirl chamber of the RWTH burner, for simplicity the swirl ratio is defined as the ratio of the tangential and axial velocity components at the inlet of the secondary stream.

Table 6.3 Burner geometry.

	Burner A	Burner Oxy-1	Burner Oxy-2
Primary stream i.d. [mm]	—	—	28
Primary stream o.d. [mm]	19	19	34
Secondary stream i.d. [mm]	30	30	45
Secondary stream o.d. [mm]	41.2	37.8	49.2
Quarl i.d. [mm]	90	90	90
Quarl height [mm]	53	53	53

Table 6.4 Summary of oxyfuel test runs.

Test	Primary Stream O_2 Vol%	Secondary + Tertiary Streams O_2 Vol%	Burner Type	Combustion Characteristics
1a	19 (rest is N_2)	Air	A	Full burnout,[a] stable swirl flame
1b	19 (rest is CO_2)	21	A	Poor burnout, lifted dark flame
2	19 (rest is CO_2)	≥ 34 (rest is CO_2)	A	Full burnout, stable swirl flame
3	19 (rest is N_2)	≥ 27 (rest is CO_2)	A	Full burnout, stable swirl flame
4	19 (rest is CO_2)	23 (rest is CO_2)	Oxy-1	Full burnout, stable swirl flame
5	19 (rest is CO_2)	21 (rest is CO_2)	Oxy-2	Full burnout, stable swirl flame
6	17 (rest is CO_2)	19 (rest is CO_2)	Oxy-2	Full burnout, stable swirl flame
7	16 (rest is CO_2)	18 (rest is CO_2)	Oxy-2	Full burnout, stable swirl flame
8	19 (rest is N_2)	Air	Oxy-2	Full burnout, stable swirl flame

[a] Full burnout means CO levels are constantly less than 50 ppm as measured at the furnace exit.

through an annulus surrounding the primary stream inlet. Global oxygen/fuel ratio was kept at 1.3, and at the local level (at the burner) this ratio was 0.6 for all cases. The details of the three different burner geometries that were tested are given in Table 6.3.

6.2.1   Development and tests of oxycoal swirl burners

In the frame of the OXYCOAL-AC project, a series of test runs was performed, providing the experimental database for further burner development (for overall conditions, see Table 6.4). The coal used was pre-dried Rhenish lignite (see Table 6.1); the thermal load was 40 kW for all tests.

First tests were performed with burner A (at left in Figure 6.3), which was designed and successfully used for pulverised coal combustion in air. Operation with air (Test 1a in Table 6.4) led to a stable, luminous, swirl flame and full burnout.

The oxyfuel test (Test 1b in Table 6.4) was carried out with 21 vol% oxygen concentration in the oxidiser (CO_2/O_2) mixture. It led to a lifted, dark flame with poor burnout and reactions somewhere in the middle of the furnace. Flame attachment and its stabilisation became possible only after increasing the O_2 concentration in the secondary and tertiary streams to levels above 34 vol% (Test 2) or by using air as the primary stream and supplying CO_2/O_2 mixture just as secondary and tertiary streams but with 27 vol% or higher O_2 content (Test 3).

6.2.2 Measures for oxycoal swirl flame stabilisation

The higher heat capacity of the gas mixture delays the heating of the pulverised coal-gas mixture, thus influencing particle devolatilisation, ignition, and combustion. This was confirmed by Test 3, demonstrating that the use of air as primary stream led to flame stabilisation at lower (27 vol%) O_2 concentrations compared to Test 2, where CO_2/O_2 was used as a primary stream. During Tests 2 and 3, a full burnout was achieved at O_2 concentrations lower than 34 and 27 vol%, respectively. However, the flame was not always attached to the burner, fluctuating between a stable and an instable state. The reasons for this behaviour can be related to the volume changes due to the endothermic gasification (Boudouard) reaction, which becomes the dominant heterogeneous reaction at particle temperatures above 1000 °C and lack of oxygen. Based on these observations during the experimental work, the following measures for oxyfuel flame stabilisation with an O_2 concentration in the CO_2/O_2 mixture equal or lower than those in air were derived:

- Compensation for higher molar heat capacity of the gas mixture by increasing the heat supply to the burner quarl
- Maintaining constant velocities at the burner by stabilising the devolatilisation rates and the CO production rates from the heterogeneous reactions (including the endothermic Boudouard reaction)

This can be obtained by:

- Strong recirculation of the hot combustion products, which will provide the necessary heat to compensate for the higher heat capacity and the enthalpy of the endothermic Boudouard reaction.
- Reducing the gas mass flow through the burner (i.e., dense flow approach) in order to increase the ratio of recirculated flue gas and incoming gas mixture. This generates under-stoichiometric conditions in the near burner region and increases the flame temperature, thus intensifying the reaction rates.

Based on these considerations, the shape of the quarl of Burner A was slightly modified (Oxy-1), as shown in the center of Figure 6.3, in order to allow stronger recirculation of the hot products to the burner quarl.

Numerical simulations of oxyfuel flames were performed with 21 vol% O_2 in the gas mixture for the burner designs considered above. The numerical model was developed for oxycoal conditions and presented by Toporov *et al.* [212, 213]. Based on these simulations, it was observed that the Oxy-1 design generates a larger and stronger internal recirculation zone compared to Burner A. The reverse flow draws hot combustion products back towards the burner inlet, providing the high heat input required to compensate for the higher heat capacity of the incoming fresh gas mixture. As a result, faster release and ignition of the volatiles as well as enhanced particle ignition and gasification were predicted for Burner Oxy-1 compared to Burner A (see Figure 6.4). Burner Oxy-1 was built and tested experimentally in Test 4 in oxyfuel conditions, with the result of being a stable flame and full burnout at an O_2 content of 23 vol% in the CO_2/O_2 mixture. Further decrease of the O_2 concentration in the burning mixture during Test 4 kept the flame stabilised at the burner, but burnout became worse. Therefore, the burner geometry was redesigned to a single annular orifice-type burner shown at right in Figure 6.3. The aerodynamics of the redesigned burner Oxy-2 differ from burners A and Oxy-1

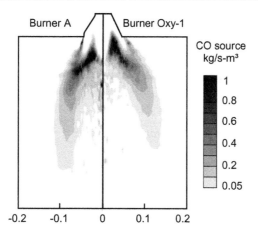

Figure 6.4 Predicted distribution of CO release rates for burner A and burner Oxy-1.
Source: Reprinted from Fuel, Vol.88/7, P. Heil, D. Toporov, H. Stadler, S. Tschunko, M. Förster, R. Kneer, Development of an oxycoal swirl burner operating at low O_2 concentrations, 1269–1274, (2009), with permission from Elsevier.

and allow further increase of the internal recirculation of hot gases, thus enabling flame stabilisation at even lower O_2 concentrations than 23 vol%. CFD was intensively used as a design tool in order to find the optimum geometry and operating conditions. The design of Burner Oxy-2 provides faster particle ignition, with predicted high CO production rates much closer to the burner inlets and as a result possible flame stabilisation near the burner compared to Oxy-1, as shown in Figure 6.5. Burner Oxy-2 was built and tested experimentally in Tests 5–7 in oxyfuel conditions. The flame remained stable

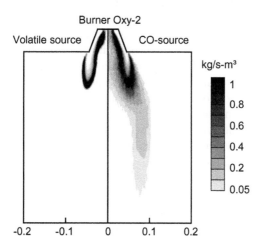

Figure 6.5 Predicted distribution for volatile- and CO release rates for burner Oxy-2.
Source: Reprinted from Fuel, Vol.88/7, P. Heil, D. Toporov, H. Stadler, S. Tschunko, M. Förster, R. Kneer, Development of an oxycoal swirl burner operating at low O_2 concentrations, 1269–1274, (2009), with permission from Elsevier.

and full burnout was kept up to an O_2 content of 18 vol%. Test 8 demonstrated the possibility of Burner Oxy-2 operating in air-combustion mode as well.

6.3   Oxycoal swirl flame properties

This section describes the properties of an oxycoal swirl flame obtained with burner Oxy-2 and through detailed in-flame measurements.

6.3.1   Measurement techniques

All measurements were conducted in the range between an axial distance of 0.025 m and 0.5 m from the burner. To investigate radial profiles of the flame, the individual measurement techniques can be traversed radially according to the furnace coordinate system.

A water-cooled suction pipe probe with a ceramic tip was used to measure gas concentrations in the flame. The probe is connected to a flue gas analysis system determining concentrations of O_2, CO_2, CO, NO, and NO_2. The concentration of oxygen was measured by a magnetomechanical analyser, and NO and NO_2 were measured by a UV photometer. CO_2 and CO were measured by a nondispersive infrared absorption. To minimise the influence of the sequence of the measurement positions, the order was randomly selected. First, the burner was moved to a randomly selected axial position. Then the radial profile of the flue gas concentrations was measured in a randomly selected order of measurement positions. The measurement time for every position was about five minutes.

The radial profiles of the oxygen concentration are the results of two measurements on two different days in two different sequences of measurement positions. The mean value for each measurement position was calculated and evaluated by variance analysis. The value of the confidence interval is equivalent to the value of two standard deviations.

A traversable IFRF-suction pyrometer was used for measurements of the gas temperature. It consists of a water-cooled stainless steel probe with a ceramic tip ($D_{outer} = 27$ mm). A PtRh/Pt thermocouple is mounted within the tip such that the thermocouple itself is shielded from the radiation of the surroundings. The gas is drawn with a velocity of about 150 m/s between the ceramic shield and the ceramic tip surrounding the thermocouple attempting to obtain an equilibrium thermocouple temperature as close as possible to the gas temperature.

Velocity measurements were carried out using laser Doppler anemometry (LDA). LDA permits nonintrusive measurements of axial and tangential velocities with high spatial and temporal resolution.

For measurements within a coal flame, no additional seeding with tracer particles was used, since the particle density due to coal and ash particles was found to be sufficient. The particle size distribution ($D_{p50} < 25$ µm) is small enough that the mean particle velocity can be considered to represent the gas velocity, as described by Jensen et al. [214].

Because of the large working distance between the optics and the measurement position within the flow, individual beam optics with single beam adjustment are used as transmitting optics. Hence, by increasing the beam distance ($s = 0.21$ m), a reasonable small measurement volume of 150 μm diameter and 2.5 mm length can be obtained. To reduce the effort of alignment, a backscattering setup is chosen where by the transmitting and receiving optics are mounted on a single frame.

The signals from the probe were analysed in a Dantec burst spectrum analyser after passing a photomultiplier. Data rates were in the range between 5 and 50 Hz for both the axial and the tangential component of the velocity. Sample times were 180 s, leading to 1000–10000 individual particles being recorded. The velocities reported here are arithmetic mean velocities of the recorded individual particles. In addition a 95 % confidence interval for the data is given.

Two-colour pyrometry was used for the determination of coal particle temperatures. The method is based on nonintrusive optical measurement of the intensity of infrared radiation at two selected wavelengths of a hot body. The investigated body—the coal particle—is treated as grey. With this assumption, the ratio of the radiation intensities at the two wavelengths is constant for a specific temperature. Moreover, the ratio is independent of the emissivity of the solid body. This means that the temperature can be determined from the ratio of the detected radiation intensities at the two wavelengths.

To allow for spatially resolved measurements, two optical beams have been used. The intersection of their optical paths forms the measurement volume. Thus it becomes possible to consider only particles that appear simultaneously on both wavelengths and therefore are inside the measurement volume and to discard all others. The angle between the two optical beams, however, is restricted by the furnace design to approximately 10°. The result is a stretched measurement volume of 5.6 mm diameter and a length of 80 mm.

6.3.2 Operating conditions

For the current investigations, the oxidiser mixture (CO_2/O_2) was provided by a gas mixing unit. The mixture was then split into several streams according to Table 6.5. The experiments were carried out at a thermal load of 40 kW. The overall oxygen/fuel ratio was 1.3, whereas the local ratio at the burner was set to 0.6, and the remaining oxidiser mixture was injected as staging fluid through the staging stream inlet. The secondary stream is highly swirled with a ratio between tangential and axial velocity components of 1.0 at the secondary stream inlet. The tertiary stream was enabled for purging purposes only.

Table 6.5 Flow parameters for 40 kW flame experiments.

	Mass Flow (kg/h)	O_2 Content (Vol.)	CO_2 Content (Vol.)	Temperature (°C)
Coal	6.5	—	—	—
Primary stream	17.6	0.19	0.81	40
Secondary stream	26.6	0.21	0.79	60
Tertiary stream	1.5	0.21	0.79	60
Staging stream	54.9	0.21	0.79	900

6.3.3   Detailed in-flame measurements

Velocity measurements

The velocities reported here are arithmetic mean velocities of the recorded individual particles.

The profiles of the measured axial and tangential velocities are shown in Figure 6.6, respectively. Clearly visible at the top of Figure 6.6 is the maximum axial velocity of 7 m/s at the plane next to the burner. This maximum corresponds to the outer diameter of the burner quarl and is due to the entry of the secondary stream. Since there is no second maximum, it can be concluded that the primary stream jet, together with the coal, merged with the secondary stream within the quarl, forming a single stream. The velocity maximum of this stream decreases and broadens with increasing distance from the burner. Two recirculation zones can be distinguished. The measured maximum

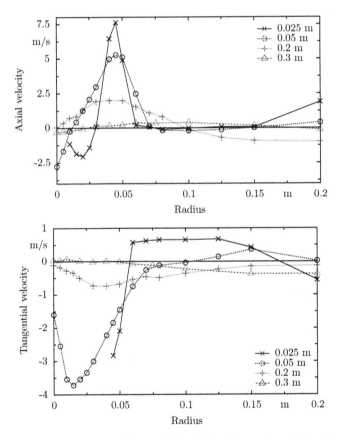

Figure 6.6 Axial velocity (top) and tangential velocity (bottom) at different axial burner distances.
Source: Reprinted from Combustion and Flame, Vol.155/4, D. Toporov, P. Bocian, P. Heil, A. Kellermann, H. Stadler, S. Tschunko, M. Förster, R. Kneer, Detailed investigation of a pulverized fuel swirl flame in CO_2/O_2 atmosphere, 605–618, (2008), with permission from Elsevier.

backward velocity inside the inner recirculation zone on the axis of the burner is almost 3 m/s. The outer recirculation zone exhibits a smaller velocity magnitude (about 1 m/s), but its volume is much larger. It reaches out from a radius of r = 0.1 m to the furnace walls. The confidence interval is well below 0.1 m/s for almost all measurement points except for those close to the furnace axis. On the axis, the data rate is rather low because of the high particle density.

From the plots of the tangential velocity (bottom of Figure 6.6) it can be seen that a high swirl is induced. This swirl is centered on the axis and hardly extends radially any further than 0.050 mm from the axis. The swirl induces a low-pressure region on the axis, which causes the strong inner recirculation.

The steep increase between a radius of 0.05 and 0.06 m at an axial distance of 0.025 m again marks the secondary air as already described for axial velocity. At this measurement plane no data are available for the area close to the axis, since the particle density is too high for individual particles to be detected. Therefore, measurements consist only of a marginal number of particles, and thus these data points are taken out of consideration. In the burner vicinity (axial distances 0.025 and, at some points, 0.05 m) the flow in the region outside a radius of 0.05 m rotates in the opposite direction as inside that radius. This behaviour is thought to originate from an interaction of the window purge flow, the tertiary air, and the burner mount, which carries the burner and functions as the top end of the furnace. At planes close to the burner, the observation window is almost halfway covered by the burner mount. Due to ash depositions, the window opening is not an ideal cylinder, giving room for a redirection of the recirculation. The window purge flow, which is slightly swirled itself, might also be significantly influenced by the burner mount and the tertiary air. Any combination of these disturbances might give rise to the change of swirl direction in the outer region at the near burner plane.

The zero crossing of the tangential velocity showed an offset to the axis for all profiles, which was also evident in the maxima of the axial velocity. This offset can be caused by a minor misalignment between the burner and the furnace axes. In addition, an offset can also be due to the alignment procedure of the optical axis of the LDA system with the burner axis.

Species concentrations and flue gas temperature measurements

The radial profiles for the gas temperatures and O_2 concentrations, obtained for five different planes, are shown in Figure 6.7. It can be observed that at an axial distance of x = 0.05 m, a significant proportion of the combustion has already occurred. The high gas temperature (around 1000 °C) and low oxygen concentration (around 2 vol% O_2), measured between 0 < r < 0.04 m, indicate that an intense combustion is taking place in the region between the internal recirculation and the incoming main jet. Three peaks can be observed in the O_2 profiles. One is obtained at r = 0.04 m and is associated with the "cold" secondary stream, resulting in a local peak in the O_2 profile reaching 5 vol% and a drop in the gas temperature at R = 0.05 m to a value of 900 °C. The local O_2 minimum obtained at r = 0.05 m (around 2 vol% O_2) can be related to the intense combustion that takes place in the zone of contact between the incoming main

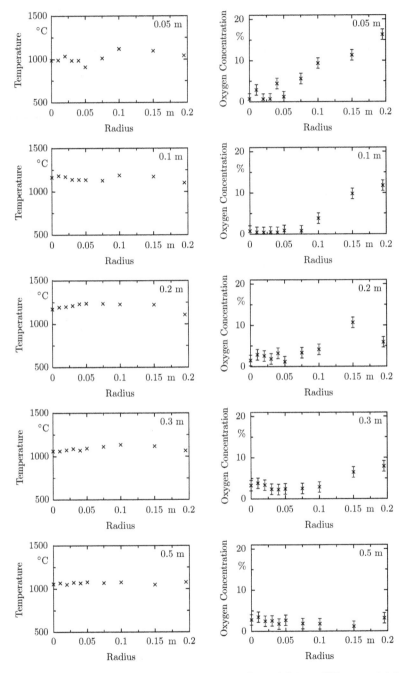

Figure 6.7 Temperature (left) and oxygen concentration (right) at different axial burner distances.
Source: Reprinted from Combustion and Flame, Vol.155/4, D. Toporov, P. Bocian, P. Heil, A. Kellermann, H. Stadler, S. Tschunko, M. Förster, R. Kneer, Detailed investigation of a pulverized fuel swirl flame in CO_2/O_2 atmosphere, 605–618, (2008), with permission from Elsevier.

primary/coal/secondary stream and a mixture of the hot externally recirculated flue gases and oxygen-rich staging fluid. The staging jet itself forms the third peak in the O_2 profile with an estimated value of about 16 vol% in the near furnace wall region.

At x = 0.1 m axial distance, the O_2 and gas temperatures profiles reveal that an intense combustion continues between 0.05 m and 0.1 m. The O_2 concentration between 0 < r < 0.75 m is almost zero as a direct influence of the low local (at burner) excess air ratio, which leads to fast consumption of the oxygen due to volatiles combustion and to an increase of the gas temperature to 1200 °C at the burner axis. With increasing axial distance, the oxygen profile becomes flatter and the region with low concentrations near the burner axis becomes broader.

At x = 0.2 m axial distance, the O_2 concentration profile has a local minimum near the furnace wall, indicating the presence of a strong external recirculation (ER) at this level. This ER pushes the staging fluid stream into the center of the furnace, thus increasing the O_2 concentrations near the burner axis.

In the area between 0.3 m and 0.5 m axial distance, the oxygen concentration stabilises at around 3 vol%, forming a flat profile with slight variations due to the heterogeneous reactions and mixing. The gas temperatures stabilise in the range between 1050 and 1100 °C.

Figure 6.8 shows the radial profiles of the nitrogen oxide concentrations. At an axial distance of 0.05 m, two peaks occur—one at the burner axis and the other at r = 0.04 m. At a burner distance of 0.1 m and radial distance between 0 < r < 0.075 m, the NO concentration reaches values of 500 ppm. Further downstream the NO concentration decreases.

Although the concentration of carbon monoxide was also measured, detailed CO profiles are not presented in the current study due to the limitations related to the

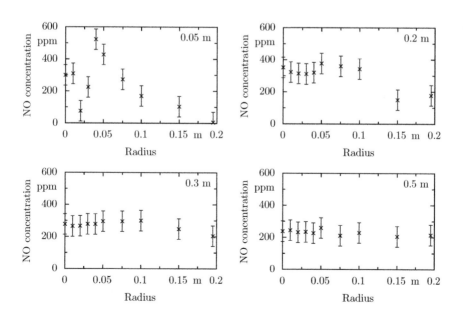

Figure 6.8 NO concentrations measured at different axial burner distances.

measurement device, which cannot capture values above 3000 ppm. The observed zone with CO levels above 3000 ppm was $x = 2D$ long and $r = 0.75D$ wide, with D being the burner quarl diameter.

Particle temperature measurements

The temperature of the coal particles was investigated over the entire radius of the combustion chamber at axial distances from 0.05 m to 0.5 m. However, at distances of 0.3 m and above, just a few particles were detected. This can be explained by the fact that the pyrometer can only detect particles that are hotter than the background and that at this distance the majority of particles have already burnt out. The particle temperature decreases with increasing axial distance from the burner, from approximately 1450 °C close to the burner to about 1100 °C at 0.2 m, whereas the radial position does not influence the temperature significantly (see Figure 6.9). The little radial changes result at least partly from the long measurement volume, which results in a floating average smoothing of the temperature. The even temperature distribution is interrupted at 0.05 m axial distance and a radial position of 0.05 m, where the cold secondary stream enters the furnace and the particle temperatures drop to 1300 °C, which presumably is still biased towards higher particle temperatures at this position due to the inability of the pyrometer to detect particles colder than the background and the above-mentioned smoothing effect.

6.3.4   Summary of the experimental results

Detailed in-flame data are provided that give a more specific insight into the underlying mechanisms characterising an industrial swirl oxyfuel flame, stabilised at 21 vol% O_2 concentration following the method described in [215]. The investigated flame is

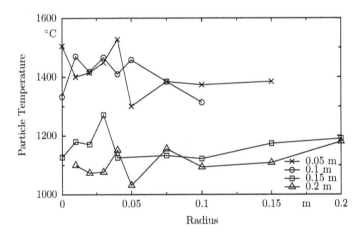

Figure 6.9 Particle temperatures measured at different axial burner distances.
Source: Reprinted from Combustion and Flame, Vol.155/4, D. Toporov, P. Bocian, P. Heil, A. Kellermann, H. Stadler, S. Tschunko, M. Förster, R. Kneer, Detailed investigation of a pulverized fuel swirl flame in CO_2/O_2 atmosphere, 605–618, (2008), with permission from Elsevier.

characterised by a strongly swirled secondary stream, which generates a large, strong internal recirculation zone with a length in axial direction of $x/D = 2$ and a maximum backward velocity reaching 3 m/s, measured on the burner axis. The burner design (larger core tube diameter and a high ratio between quarl length and diameter) creates a highly mixed region inside the burner quarl, thus allowing the hot recirculated gases to enter closer to the burner. The reverse flow draws hot combustion products with a temperature of about 1200 °C back towards the burner inlet, providing the high heat input required for fast particle heating as well as to compensate for the higher molar heat capacity of the incoming fresh gas mixture and for the endothermic Boudouard reaction. The combined influence of the annular jet, the strong, well-defined, and enclosed internal recirculation zone (IRZ), and the high swirl of the secondary stream result in the coal particles being subjected to high radial forces. Just a few particles have sufficient axial momentum to partially penetrate into the IRZ and to burn there. Instead, the majority are confined to merging with the secondary stream, thus forming a main coal/primary/secondary stream that enters the combustion chamber at the outer diameter of the burner quarl. The low excess air ratio at the burner increases the ratio between the recirculated mass and the incoming primary/secondary stream, thus improving the heating of the incoming coal containing CO_2/O_2 flow with the hot recirculation. The particles ignite in the oxygen-rich corridor inside the quarl between the outer boundary of the hot IRZ and the incoming main stream. Thus, fast release and ignition of the volatiles as well as an improved particle ignition and gasification are achieved, and as a result a stable swirl oxyfuel flame is obtained. The under-stoichiometric conditions at the burner lead to an incomplete combustion in the near burner region (NBR), thus favouring the Boudouard gasification reaction. As a result, a large zone of high CO concentration and nearly zero O_2 concentration are formed around the burner axis. The fuel nitrogen is mainly released during devolatilisation in the NBR, and the subsequent high emission values (500 ppm NO at 0.1 m) are mainly due to this volatile nitrogen conversion to NO. The rest of the volatiles, char, and CO burn completely in the outer region of the main coal/primary/secondary stream and downstream of the flame, where the unburnt fuel mixes with the hot mixture of externally recirculated combustion products and the oxygen-rich staging fluid.

6.3.5 Validation of the numerical model

The numerical model

A model for numerical simulation of pulverised coal combustion under oxyfuel conditions was developed using the CFD code FLUENT 6.2 with modified (UDF-based) devolatilisation, char oxidation, and gas phase combustion submodels considered in detail in Chapters 4.2 and Chapter 4.1.2 respectively. The flow field was solved using the standard k-ε turbulence model with the SIMPLEC algorithm for velocity-pressure coupling. Transport equations for the mass fractions of seven different gas species (volatiles ($C_xH_yO_lS_nN_m$), water vapour, carbon monoxide, nitrogen, sulphur dioxide, oxygen, and hydrogen) were calculated, with source terms defined according to the reaction scheme proposed below. The carbon dioxide mass fraction was calculated from mass balance, being the most abundant species in the gas mixture. The following kinetic mechanism was assumed for the combustion of volatiles:

$$C_xH_yO_lS_nN_m + \left(\frac{x}{2} + n - \frac{l}{2}\right)O_2 \rightarrow xCO + \frac{y}{2}H_2 + nSO_2 + \frac{m}{2}N_2 \qquad (6.1)$$

$$CO + \frac{1}{2}O_2 \rightarrow CO_2 \qquad (6.2)$$

$$H_2 + \frac{1}{2}O_2 \rightarrow H_2O \qquad (6.3)$$

The interaction between the turbulence and chemical reactions in the gas phase was modelled using the finite rate/eddy dissipation model, with kinetic rates taken from Shaw *et al.* [216] and from Dryer and Glassman [217] for the volatiles (Eq. 6.1) and carbon monoxide (Eq. 6.2) oxidation reactions, respectively. The oxidation of hydrogen (Eq. 6.3) was assumed to be infinitely fast.

Coal particle combustion was simulated based on a stochastic Lagrangian procedure to track the particle trajectories in the flow field. The coal particle undergoes decomposition into char and volatile material, the former burning slowly in the later stages of the flame, forming CO for pulverised coal combustion conditions, whereas the volatile material is assumed to rapidly form CO and, subsequently, CO_2. The devolatilisation process is modelled using the chemical percolation devolatilisation (CPD) model [64]. The input parameters for the CPD model are based on the ultimate analysis of the coal, thus making it sensitive to the investigated coal type. Three main heterogeneous reactions during coal char combustion are considered:

$$C_{char} + \frac{1}{2}O_2 \rightarrow CO \qquad (6.4)$$

$$C_{char} + CO_2 \rightarrow 2CO \qquad (6.5)$$

$$C_{char} + H_2O \rightarrow CO + H_2 \qquad (6.6)$$

The char burnout rates are obtained from an apparent surface kinetic and diffusion mass transfer rates. A first-order apparent kinetic rate model was applied for reactions (Eqs. (6.4)–(6.6)) following the kinetic rates provided in Tables 4.8, 4.9, and 4.13, respectively.

The processes of devolatilisation and char oxidation were assumed to happen in parallel, governed by particle temperature only, thus making an overlap possible. The three heterogeneous reactions were also assumed to run simultaneously.

The radiative heat source was calculated as a function of the local irradiation, calculated by the discrete ordinates radiation model. The local absorption coefficient was calculated as the sum of the particle and gas absorption coefficients. The latter was based on the weighted sum of grey gases model, which considers the absorption coefficients of different grey gases of the mixture. Although the WSGGM model is not fully applicable for oxyfuel conditions (see Chapter 4.3), the furnace size is small enough (short pathlengths) to distinguish any differences in the gas emissivity.

NO concentrations were calculated (as a first step) using standard FLUENT postprocessing NO_x models, including thermal, prompt, and fuel NO production as well as NO consumption due to reburning. Eighty percent of the fuel nitrogen was assumed to be released from the volatiles due to the coal type and the high pyrolysis temperature. O- and OH-radical concentrations were estimated using a partial equilibrium approach.

The boundary conditions are the same as given in Table 6.5. The computational domain is three dimensional and represents one-sixth of the furnace volume, applying periodic boundary conditions. The numerical grid contains 590,800 cells. The temperatures and emissivities of the furnace and burner walls are taken to be 1000 °C ($\varepsilon = 0.7$) and 300 °C ($\varepsilon = 0.2$), respectively.

Numerical results

A comparison between experimental data and numerical simulations was carried out. Figure 6.10 shows the axial velocity profiles of the flame at 0.05 m and 0.2 m axial distance to the burner. At the position close to the burner, the axial velocity maximum of experiment and simulation are in good agreement with respect to radial position and velocity magnitude. The width of the inner recirculation zone, however, is over-predicted, and the calculated maximum value of the back-flow velocity is lower than the measured one. At the outer wall of the furnace, the calculation shows a local peak that is not observed in the measurements. This peak is caused by the staging air, which, according to the measurement, seems to be already dissipated at this level. At 0.2 m burner distance, the calculated velocity magnitude is in agreement with the experimental values, although the profile is slightly shifted in a radial direction, thus predicting a wider flame. This can explain why the back-flow observed in the experiment is not predicted at this burner distance.

Figure 6.11 shows the profiles of the tangential velocity calculated for 0.05 m and 0.2 m axial distance. The tangential velocity maximum at $x = 0.05$ m is slightly under-predicted; the maximum obtained at $x = 0.2$ m is strongly overpredicted. In both cases the radial position of the velocity maximum is shifted slightly in radial direction in comparison to the experiment.

/The oxygen concentration profiles are shown in Figure 6.12. It can be observed that in the near burner axis region the predicted and measured oxygen concentrations for both profiles are very low. This is a direct result of the burner operating conditions

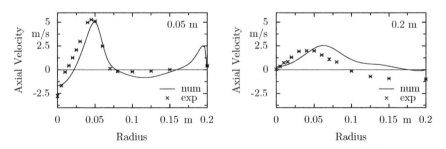

Figure 6.10 Axial velocities at 0.05 m (left) and 0.2 m (right) axial burner distance.
Source: Reprinted from Combustion and Flame, Vol.155/4, D. Toporov, P. Bocian, P. Heil, A. Kellermann, H. Stadler, S. Tschunko, M. Förster, R. Kneer, Detailed investigation of a pulverized fuel swirl flame in CO_2/O_2 atmosphere, 605–618, (2008), with permission from Elsevier.

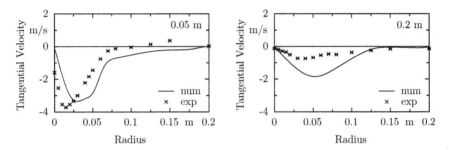

Figure 6.11 Tangential velocities at 0.05 m (left) and 0.2 m (right) axial burner distance.
Source: Reprinted from Combustion and Flame, Vol.155/4, D. Toporov, P. Bocian, P. Heil, A. Kellermann, H. Stadler, S. Tschunko, M. Förster, R. Kneer, Detailed investigation of a pulverized fuel swirl flame in CO_2/O_2 atmosphere, 605–618, (2008), with permission from Elsevier.

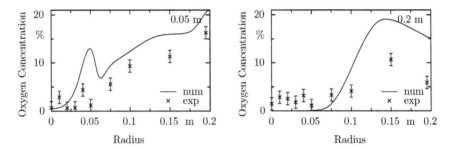

Figure 6.12 Oxygen concentration at 0.05 m (left) and 0.2 m (right) axial burner distance.
Source: Reprinted from Combustion and Flame, Vol.155/4, D. Toporov, P. Bocian, P. Heil, A. Kellermann, H. Stadler, S. Tschunko, M. Förster, R. Kneer, Detailed investigation of a pulverized fuel swirl flame in CO_2/O_2 atmosphere, 605–618, (2008), with permission from Elsevier.

with local oxygen to fuel ratio equal 0.6 and the high swirl ratio. The local maximum, which is observed at 0.05 m and is related to the secondary stream, is overpredicted and shifted in radial direction. The minimum in the oxygen profile surrounding the secondary stream is caused by the combustion taking place there due to the contact between the surrounding hot gases and the main primary/coal/secondary stream. Higher oxygen concentrations are predicted for the outer regions of the furnace compared to the measurements for both planes. This is caused by the different behaviour of the staging stream in predictions and experiment.

The calculated gas temperature profile shows steep gradients, whereas the experimental profile is flatter. Four different extrema can be observed in the gas temperature profile obtained numerically for an axial distance of 0.05 m (see Figure 6.13). The first one is located on the burner axis and is a result of the combustion of mainly volatiles, which takes place there. The second one is related to the secondary stream. The difference between the values obtained from simulation and experiments for this region can be explained by the coarser spatial resolution of the measurements performed. The same can be applied for the third peak obtained by simulation, which is caused by the combustion at the outer surface of the main primary/coal/secondary stream and is not

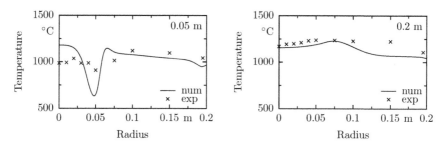

Figure 6.13 Gas temperature at 0.05 m (left) and 0.2 m (right) axial burner distance.
Source: Reprinted from Combustion and Flame, Vol.155/4, D. Toporov, P. Bocian, P. Heil, A. Kellermann, H. Stadler, S. Tschunko, M. Förster, R. Kneer, Detailed investigation of a pulverized fuel swirl flame in CO_2/O_2 atmosphere, 605–618, (2008), with permission from Elsevier.

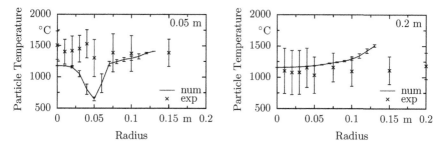

Figure 6.14 Particle temperature at 0.05 m (left) and 0.2 m (right) axial burner distance.
Source: Reprinted from Combustion and Flame, Vol.155/4, D. Toporov, P. Bocian, P. Heil, A. Kellermann, H. Stadler, S. Tschunko, M. Förster, R. Kneer, Detailed investigation of a pulverized fuel swirl flame in CO_2/O_2 atmosphere, 605–618, (2008), with permission from Elsevier.

observed in the gas temperature measurements. In the near wall region, a forth peak caused by the staging stream is predicted. At an axial distance of 0.2 m, the predicted gas-temperature profile follows the measured one.

The predicted and measured particle temperature profiles obtained for 0.05 m and 0.2 m axial distance are shown in Figure 6.14. The calculated particle temperatures follow the predicted gas temperature profile obtained for $x = 0.05$ m and are lower than the measured ones. A possible explanation might be the inability of the two-colour pyrometer used to detect particles with temperatures lower than the background temperature, which is around 940 °C at $x = 0.05$ m and 1050 °C at $x = 0.2$ m. Nevertheless, the profile shows a minimum located at $r = 0.05$ m, which can be associated with the "cold" secondary stream. At $x = 0.2$ m, the predicted temperature is higher than the measured one. This difference can be explained by the fact that in the numerical calculation, the completely burnt particles (ash particles) are removed in order to reduce the calculation time. Thus, at $X = 0.2$ m only big particles that are still burning are calculated. In reality, however, the hot ash particles are not removed from the furnace and are detected by the pyrometer. Thus, being hot but colder than the burning particles, they reduce the average measured particle temperature.

A comparison between the predicted and measured NO concentration profiles obtained for 0.05 m and 0.2 m axial distance is shown in Figure 6.15. At $x = 0.05$ m,

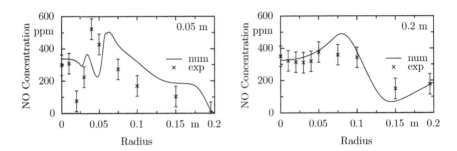

Figure 6.15 NO concentration at 0.05 m (left) and 0.2 m (right) axial burner distance.

the calculated profile follows the measured data. However, due to the predicted wider flame, it is slightly shifted radially. A similar behaviour can be seen at $x = 0.2$ m. The maximum values calculated for both axial distances can be related to char nitrogen, which is released in the hot region surrounding the main jet.

6.4 Burner scale-up

From a theoretical point of view, scaling of flames requires the simultaneous consideration of fluid flow, heat transfer, and chemical reactions in close proximity to the burner. This involves the derivation of dimensionless numbers from conservation equations of mass, energy, and momentum. However, such scaling is fraught with difficulties because not all the physical and chemical processes scale similarly. This is why in practice only a few of the scaling rules are obeyed, leading to so-called "partial modelling" or "scaling."

The burner thermal input $Q^{\cdot}$, [kW] can be expressed as follows [218]:

$$Q_0^{\cdot} = K \rho_0 U_0 D_0^2 \tag{6.7}$$

where ρ_0, U_0, and D_0^2 are the inlet fluid density, characteristic burner (gas) velocity, and burner characteristic diameter, respectively, and K is a proportional constant. When a burner is scaled from a baseline thermal output Q_0 to a different thermal input Q_{scaled} using the constant residence time-scaling approach proposed by Weber [219], the characteristic burner diameter D_{scaled} will be:

$$D_{\text{scaled}} = D_0 \left(\frac{Q_{\text{scaled}}}{Q_0} \right)^{0.33} \tag{6.8}$$

Thus, if values of D_0 and Q_0 are known, values for the scaled burner can be derived. This method assumes that the ratio D_0/U_0 remains constant while changing the burner thermal input. The ratio represents the inertial time scale of the flow, often called the convective time scale. This approach maintains the gas or particle residence times in the near burner region similarly between the small-scale system and the larger system. It implies that for a larger system with a larger combustion chamber or furnace volume, the velocities will also be higher.

Table 6.6 Burner dimensions.

	Burner Oxy-2	Burner Oxy-3
Primary stream PS i.d. [mm]	28	37
Primary stream PS o.d. [mm]	34	45
Secondary stream SS i.d. [mm]	45	57
Secondary stream SS o.d. [mm]	49.2	64
Quarl D. [mm]	90	120
Quarl height L [mm]	53	53
Tertiary stream TS-I i.d. [mm]	96	146
Tertiary stream TS-I o.d. [mm]	100	158
Tertiary stream TS-II i.d. [mm]	390	390
Tertiary stream TS-II o.d. [mm]	400	400

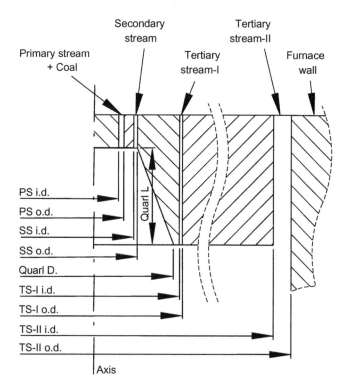

Figure 6.16 Burner geometry and dimensions.

Following this approach, Burner Oxy-2 was scaled up from 40 to 100 kW_{th} (burner Oxy-3), leading to dimensions given in Table 6.6 and the geometry given in Figure 6.16.

Numerical simulations of Burner Oxy-3 firing pulverised pre-dried Rhenish lignite were performed. The numerical model developed for PF oxyfuel combustion and

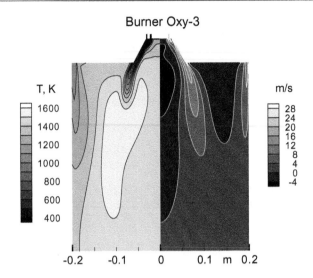

Figure 6.17 Predicted distribution of gas temperature (left) and axial velocity component (right) for the Oxy-3 burner at 100 kW.

described in detail by Toporov *et al.* [212] was upgraded by a nongrey implementation of the exponential wideband model for the local absorption coefficient according to Erfurth *et al.* [123]. The boundary conditions at the burner inlets correspond to wet recycling conditions, with an oxidiser ratio of 0.6 at the burner and 1.3 in total. The oxygen concentration in the oxidiser ($O_2/CO_2/H_2O$) mixture was set to 21 vol%.

The predicted velocity and temperature fields in the near burner region are given in Figure 6.17. As shown, the strong swirl ratio and large inner diameter of the primary stream inlet create a strong inner recirculation zone that draws hot combustion products with a temperature of about 1000 °C back towards the burner inlet, providing the high heat input required for fast particle heating as well as compensating for the higher c_p of the incoming fresh gas mixture and for the endothermic Boudouard reaction. Thus, fast release and ignition of the volatiles as well as improved particle ignition, as shown in Figure 6.18, are achieved, and as a result a stable swirl oxyfuel flame at an oxygen concentration similar to that in air can be expected.

The burner was built and tested at a load of 80 kW. Visual inspection and global measurements confirmed flame stabilisation at the burner quarl under oxycoal conditions with 21 vol% O_2 in the O_2/CO_2 mixture (dry recycle). Furthermore, the burner was successfully tested in both oxycoal conditions using RFG (wet recycling) and in conventional air combustion. The tests demonstrated the burner's ability to form a stable swirl flame and to provide good burnout in all three combustion modes, namely, in air, in O_2/CO_2, and in O_2/RFG, as shown in Figure 6.19. Successful scale-up of the burner demonstrated the viability of the measures for oxycoal swirl flame stabilisation formulated in Section 6.2.2.

Direct scale-up of the burner described above to an industrial scale oxy-firing burner in the range of several MWth, however, is not possible. Instead, a modification of a conventional air-firing utility-scale burner according to the measures for oxycoal

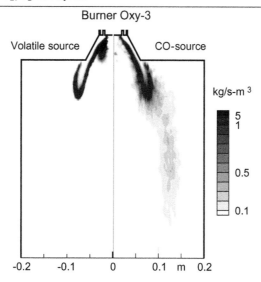

Figure 6.18 Predicted distribution for volatile and CO release rates for Burner Oxy-3.

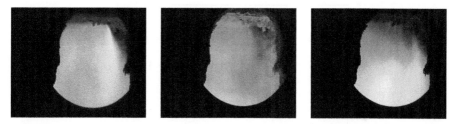

Figure 6.19 Eighty kW pulverised coal flames obtained at the test facility at RWTH Aachen for: air-firing (left) and for oxy-firing with 21 vol% O_2 content in O_2/CO_2 mixture (middle) and in O_2/RFG mixture (right). The pictures were taken from video showing a stable flame that was achieved at the burner quarl.

swirl flame stabilisation described in Section 6.2.2 and by Förster *et al.* [215] is recommended. Following this approach, the numerical model developed for oxycoal combustion was used for CFD-based design of an industrial-scale 70 MWth swirl burner. The new burner is able to operate in air-firing as well as at a wide range of O_2 concentrations in oxyfuel combustion conditions as reported by Erfurth *et al.* [213].

6.4.1 Burner aerodynamics

Cold flow validation

LDA measurements of the flow field in the near burner zone were carried out to obtain data for the overall flow pattern, including areas of flow recirculation and areas of high turbulence. The obtained data were used for validation of boundary conditions and turbulence models for burner simulations. For this reason, inert particles made of

Table 6.7 Flow parameters for the cold flow experiment.

	Flow Rate (m^3/h)	Temperature (°C)
Primary stream	10.9	20
Secondary stream	47.5	20
Tertiary stream I	2.8	20
Tertiary stream II	36.8	20

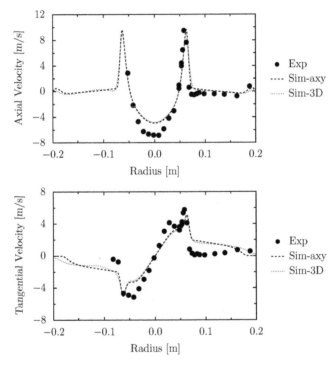

Figure 6.20 Predicted and measured profiles of axial (top) and tangential (below) velocity components at 15 mm axial burner distance. Sim-axy and sim-3D stand for axisymmetrical and three-dimensional simulations, respectively.

Al$_2$O$_3$ with a density of less than 4 g/m^3 and diameter of $D_{90} = 2$ μm were used. The particles were supplied to the burner with a mass flow rate of 8.3 × 10^{-6} kg/h. Measurements were carried out at 15 mm and 30 mm axial distance from the burner. The air mass flow was set equal to an air flame with about 80 kW thermal power. The burner settings used are shown in Table 6.7.

The measured profiles of axial and tangential components of the gas velocity are shown in Figures 6.20 and 6.21 for 15 mm and 30 mm axial distance, respectively. The burner axis is at 0 mm. Positive velocities represent top-down flow. The swirl is counterclockwise-viewed from the top of the furnace.

The maximum velocity corresponds to the outer diameter of the burner quarl and is due to the entry of the secondary stream. Since there is no second maximum, it

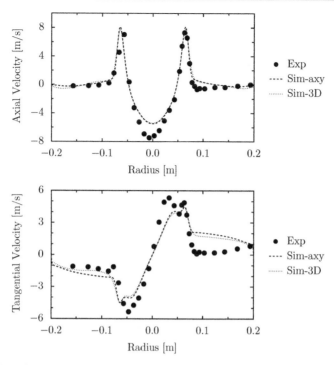

Figure 6.21 Predicted and measured profiles of axial (top) and tangential (below) velocity components at 30 mm axial burner distance. Sim-axy and sim-3D stand for axisymmetrical and three-dimensional simulations, respectively.

can be concluded that the primary stream jet carrying the particles merged with the secondary stream within the quarl, forming a single particle/primary stream/secondary stream (main jet). The maximum velocity of this stream decreases and the velocity distribution broadens with increasing distance from the burner. Two recirculation zones can be distinguished. The measured maximum backward velocity inside the inner recirculation zone on the axis of the burner is about 7 m/s. The outer recirculation zone is much slower (only about 1 m/s), but its volume is much larger. It extends from a radial position of 0.1 m out to the furnace walls. There is a local maximum of the axial velocity, located close to the furnace wall observed in the profile at 15 mm axial distance, indicating the jet coming from the tertiary stream II. At 30 mm axial distance, however, this stream is already mixed with the external backflow.

The tangential velocity profile has two maximums. The outer one corresponds with the outer diameter of the burner quarl and is due to the entry of the highly swirled secondary stream. The inner one is located inside the internal recirculation zone and is due to the fluid particles originating from the secondary stream and carrying its tangential momentum. These fluid particles are entrained into the backflow stream that is composed mainly from flue gas coming from the inner part of the furnace and that flow back to the burner inlet, driven by the pressure difference created inside the burner quarl. The minimum between the two peaks in the tangential velocity profiles corresponds to the place where the axial velocity changes its direction (zero magnitude). Further downstream, the inner peak becomes higher than the outer peak (at 30 mm axial distance) due to the smaller radius of rotation of the fluid particles that have still the same tangential momentum as those in the secondary (outer) stream.

The numerical predictions show almost no difference between 2D axysimetrical and 3D approximations. Generally, the predicted profiles of the axial velocity follow the measured ones. However, a difference is observed in reproducing the maximum of the backflow velocity. The simulations under-predict the observed maximum.

Related to the tangential velocity, the predictions provide a good qualitative agreement with the experimental data. However, the two peaks are not defined as clearly as in the measurements, and maximum values are slightly under-predicted. The values calculated for the outer region of the main jet are higher than the measured ones, with the 3D results being closer to the measurements, as shown on the left side of Figure 6.21.

Further more, a study on the influence of the turbulence model on the quality of the predictions has been performed. Results from this study, obtained for 30 mm axial distance with $k - \epsilon$ (standard), $k - \epsilon$ (renormalised group) and Reynolds stress model, are given in Figure 6.22. As shown, none of the used models could lead to significant improvement of the quality of the predictions.

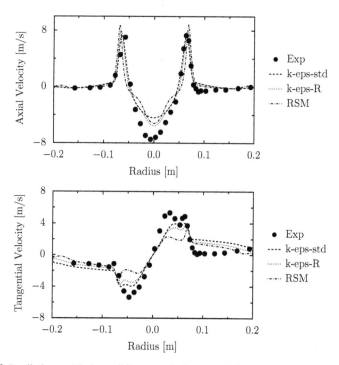

Figure 6.22 Predictions with three different turbulence models and measured profiles of axial (top) and tangential (below) velocity components at 30 mm axial burner distance. k-eps, k-eps-R, and RSM stand for $k - \epsilon$ (standard), $k - \epsilon$ (renormalised group), and Reynolds stress model, respectively.

Table 6.8 Flow parameters for 80 kW flame experiments.

	Mass Flow (kg/h)	O_2 Content (Vol.)	CO_2 Content (Vol.)	Temperature (°C)
Coal	13.0	—	—	—
Primary stream	20.63	0.19	0.81	40
Secondary stream	84.6	0.21	0.79	60
Tertiary stream	4.46	0.21	0.79	60
Staging stream	74.52	0.21	0.79	900

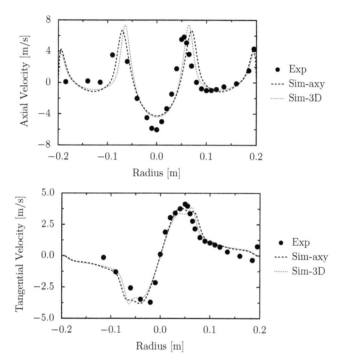

Figure 6.23 Predicted and measured profiles of axial (top) and tangential (below) velocity components at 50 mm axial burner distance in 80 kW oxycoal flame. Sim-axy and sim-3D stand for axisymmetrical and three-dimensional simulations, respectively.

Flame validations

A similar study validated the numerical data against in-flame measurements. The boundary conditions are given in Table 6.8.

The predicted and measured velocity profiles obtained for 50 mm axial distance are shown in Figure 6.23.

The comparison between the numerical simulations and the measurements show that reliable flow-field predictions of highly swirled turbulent flows (type-2 swirl flames) can be obtained using the $k - \epsilon$ turbulence model. The computations should be performed using at least a second-order numerical scheme and a fine resolution of the numerical grid. However, the results should be interpreted while taking into consideration the typical drawbacks of the $k - \epsilon$ model, as described in Section 4.1.

6.5 Summary of Chapter 6

The combustion of pulverised coal in O_2/CO_2 atmosphere leads to modified distributions of temperature, species, and radiation fluxes inside the combustion chamber due to the changed thermo-physical and optical properties compared to air. Using a burner design that was optimised for coal combustion in air will lead to flame instability and poor burnout for oxycoal combustion.

Therefore, the performance of two different burner designs, single central orifice (SCO) and single annular orifice (SAO), was examined under oxycoal conditions in a downfired oxycoal pilot plant at RWTH Aachen University. Based on CFD predictions, the main parameters influencing the stability of a O_2/CO_2 pulverised coal swirl flame were identified and investigated.

Measures for stabilisation of an oxycoal swirl flame with O_2 contents equal to and lower than those in air were derived regarding compensation of the effects of higher molar heat capacity, heat consumption, and volume changes due to gasification reactions.

The oxycoal flame was experimentally stabilised at the burner quarl by increasing O_2 concentration above 34 vol% without changes to the air-firing burner design and by modification of the burner geometry, thus changing the aerodynamics.

Thus, a novel fluid dynamics-based approach for oxycoal swirl flame stabilisation is introduced based on a thorough understanding of the underlying interactions between burner aerodynamics and reaction kinetics.

The modification of the burner allowed a decrease of the O_2 concentrations to 23 vol% for the SCO burner and to less than 19 vol% for the SAO burner. The SAO burner is characterised by a strong internal recirculation zone, which draws back hot combustion products with a temperature of about 1200 °C towards the burner inlet. This provides the heat input required for compensation of the higher c_p of the incoming fresh gas mixture and the endothermic Boudouard reaction. Thus, fast particle heating, drying, release, and ignition of the volatiles as well as an improved particle ignition and CO_2 gasification of particles are achieved. As a result, the combustion rate in the near burner region is enhanced, thereby enabling the stabilisation and full burnout of an oxycoal swirl flame with oxidiser O_2 concentrations equal to or less than those in air. Based on the accumulated experience , the SAO burner was successfully scaled up to 100 kW according to the constant residence time-scaling approach, thus demonstrating the viability of this approach. The burner was built and tested at a load of 80 kW. Visual inspection and global measurements confirmed flame stabilisation at the burner quarl under oxycoal conditions, with 21 vol% O_2 in the O_2/CO_2 mixture. Furthermore, the burner was successfully tested in both oxycoal conditions using RFG (wet recycling) and in conventianal air combustion. The tests demonstrated the burner's ability to form a stable swirl flame and to provide good burnout in all three combustion modes, namely, in air, in O_2/CO_2, and in O_2/RFG. With this status reached, the test plant in Aachen is the first plant where coal can be burned in a stable swirl flame in a CO_2 atmosphere with an oxygen content between as low as 18% up to above 30%. A scale-up of the burner described above to an industrial scale oxy-firing burner in the range of several MW_{th}, however, is not recommended. Instead, a modification of a conventional air-firing utility scale burner according to the measures for oxycoal swirl flame stabilisation described in Section 6.2.2 is suggested.

7 NO$_x$ Emissions During Oxycoal Combustion

This chapter presents the results of experimental investigations on NO$_x$ emissions from coal combustion in the oxycoal pilot plant that was described in Section 6.1. The impact of five operational parameters has been studied. The main idea of considering the first group of parameters, namely, (1) the burner excess oxygen ratio λ_b, and (2) the oxygen concentration, focuses on describing the chemical effects in NO$_x$ conversion. Considering the second group of parameters, namely, (3) the oxidiser stream inlet temperature; (4) the secondary stream momentum; and (5) the primary stream momentum, addresses the effect of the burner aerodynamics on NO$_x$ conversion.

7.1 Test conditions

Three different PF combustion modes were considered:

- Firing in air (called air mode)
- Firing in a mixture of oxygen and carbon dioxide (called O$_2$/CO$_2$ mode) by varying O$_2$ concentration in the O$_2$/CO$_2$ mixture
- Firing in a mixture of oxygen and recycled wet flue gas (RFG) (called O$_2$/RFG mode)

The conditions for the experiments are summarised in Table 7.1.

The burner air ratio was varied in a range of $0.42 \leq \lambda_b \leq 1.05$. The oxygen concentration in the oxidiser has also been varied in the range from 18 to 27 vol%.

To eliminate aerodynamic effects, for the first two parameters, the gas streams supplied to the burner (coal carrier stream, oxidiser, and secondary oxidiser stream as well as the inner inertisation) were kept at a constant volume flow throughout all measurements. To allow for a variation of the burner air ratio λ_b while keeping aerodynamic conditions constant, the coal mass flow and thus the thermal load had to be adjusted. While the local air ratio was altered (see Figure 7.1), the thermal load of the swirl burner varied from 54 to 100 kW$_{th}$ in air, from 40 to 74 kW$_{th}$ in O$_2$/CO$_2$, and from 70 to 74 kW$_{th}$ in O$_2$/RFG mode. During the changes of the oxygen concentration in the oxidiser (see Figure 7.2), the thermal load in O$_2$/RFG mode ranged from 60 to 70 kW$_{th}$, whereas in O$_2$/CO$_2$ mode it varied from 47 to 87 kW$_{th}$ with O$_2$/CO$_2$ as coal carrier and from 52 to 65 kW$_{th}$ with CO$_2$ as coal carrier (primary) stream.

Combustion of Pulverised Coal in a Mixture of Oxygen and Recycled Flue Gas. http://dx.doi.org/10.1016/B978-0-08-099998-2.00007-2

Table 7.1 Operational parameters for NO emission tests at oxycoal pilot plant at RWTH Aachen University. RR stands for recycle ratio.

| Mode | λ_b | O₂ Concentration | | Thermal Input (kW) | Volume Flows | | | RR |
		Oxidiser (Vol%)	PS (Vol%)		PS (m^3/h)	SS (m^3/h)	TS-II (m^3/h)	
Air	0.42–0.77	21	19	100–54	9.4	24.5	72–21	
O₂/CO₂	0.57–1.05	21	19	74–40	9.4	24.5	43–6	
O₂/RFG	0.60–0.57	21	19	74–70	9.4	24.5	43–39	0.58–0.59
O₂/CO₂	0.6	18–27	16–19	47–87	9.4–8.2	24.5	39–37	
O₂/CO₂	0.6	20–25	0	52–65	9.4	24.5	30	
O₂/RFG	0.6	18–21	19	60–70	9.4	24.5	39	0.58–0.60

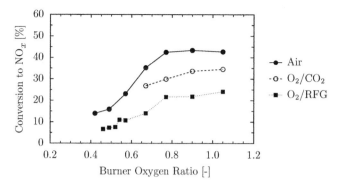

Figure 7.1 Conversion of fuel N as a function of burner excess oxygen ratio.

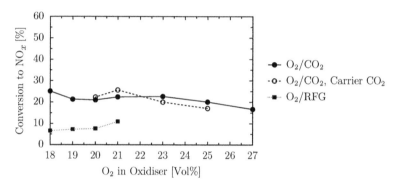

Figure 7.2 Conversion of fuel N as a function of oxygen concentration.

The staging stream was adapted accordingly to retain the overall oxygen ratio and to compensate for changes of the oxidiser oxygen concentration. Therefore, a lower burner oxygen ratio (given an unchanged oxygen content) resulted in a higher coal mass flow and a larger staging stream, whereas a lower oxygen content (with λ_b = const.) decreased the coal mass flow and left the volume flow of the staging stream constant.

The overall oxygen ratio was set as constant at λ = 1.2. The wall temperature was kept within a range between 900 °C and 950 °C.

The NO concentration was measured by means of nondispersive ultraviolet spectrography (NDUV) in dry flue gas in the stack. The measurement range was from 0 to 1000 ppm, with a maximum relative error of 0.5% of the upper limit. For comparison of emissions from the individual operation modes, the NO concentration in the flue gas has been converted to emissions based on the thermal input. For this, NO$_x$ is considered as NO$_2$.

7.2 Influence of the excess oxygen ratio at the burner

The NO$_x$ emission characteristics of the three cases, (1) air mode; (2) O$_2$/CO$_2$ mode; and (3) O$_2$/RFG mode, were evaluated among several conditions of the excess oxygen ratio at the burner. The oxygen concentration at burner inlet was set to 21 vol% for all of the combustion modes. The flow rate of the gas mixture was kept constant. The excess oxygen ratio at the burner was varied by controlling the coal feed rate. The excess oxygen ratio λ is defined as follows:

$$\lambda = \frac{m_{O_2}/m_{fuel}}{(m_{O_2}/m_{fuel})_{stoichiometric}} \tag{7.1}$$

Here the ratio m_{O_2}/m_{fuel} gives the mole ratio of oxygen to fuel. The conversion ratio (CR in %) of fuel-N to NO_x is calculated as follows:

$$CR = NO_{x_{stack}}/NO_{x_{total}} \times 100 \tag{7.2}$$

where $NO_{x_{stack}}$ is actual NO_x measured at the stack in mg/MJ (containing the fuel NO_x and thermal NO_x), and $NO_{x_{total}}$ is total NO_x, which theoretically could be emitted when all fuel N was converted to NO_x, in mg/MJ. Figure 7.1 shows the influence of the excess oxygen ratio at the burner upon conversion to NO_x during the three different operation modes. There is an obvious trend of increase in NO_x emissions with increase of the burner excess oxygen ratio. This trend is valid for all three combustion modes.

In air mode, the NO_x emissions with this burner are significantly higher than emissions obtained with state-of-the-art low-NO_x burners. The reason for this behaviour is twofold. First, the walls of the test facility are heated in order to obtain an adiabatic furnace so that NO_x emissions cannot be simply compared to an industrial application. Second, the burner has been optimised for oxyfuel combustion by inducing a strong inner recirculation of hot flue gas, which is necessary to ignite the reactants under oxyfuel conditions (Type 2). This causes high temperatures in the burner vicinity and thus favours the formation of thermal NO.

In O_2/CO_2 mode, the conversion to NO_x is about 25% to 30% lower than in air mode.

For the O_2/RFG mode, the conversion of fuel N is much lower compared to other combustion modes, reaching a reduction by approximately 50% compared to O_2/CO_2 mode. This is mainly due to the fact that NO_x contained in RFG is supplied by the secondary stream back to the flame, and, thus, it is destructed by reducing conditions due to volatile matter.

The general trend to increase NO_x emissions with increased burner oxygen ratio in oxyfuel combustion shown by Kiga et al. [220] for an oxidiser oxygen concentration of 30 vol% can be confirmed with the current experiments for an oxygen concentration of 21 vol%.

7.3   Influence of oxidiser O_2 concentration on NO_x emissions

One important design parameter in oxyfuel combustion is the O_2 concentration in the oxidiser. The burner has to operate with an oxygen concentration that is determined from the desired temperature and heat transfer in the furnace. Thus, a further investigation of NO_x emissions with respect to the O_2 concentration in the oxidiser has been carried out. Figure 7.2 presents data obtained for O_2/CO_2 and O_2/RFG combustion modes.

7.3.1 Influence of oxidiser O$_2$ concentration in the primary stream

An O$_2$/CO$_2$ combustion mode with varied O$_2$ concentration in the coal carrier stream has been investigated. The NO$_x$ emission characteristics in the case of the four levels of O$_2$ concentrations in oxidised gas are evaluated at the same excess oxygen ratio at the burner. It was found that the conversion to NO$_x$ is almost the same value, despite the existence of O$_2$ in the primary stream. This is due to the diffusion of pulverised coal particles at the combustion chamber, as discussed in Chapter 6. Pulverised coal is mainly fired after mixing with the secondary stream. It seemed that the influence of the existence of O$_2$ in the primary stream is negligible because the volume of the secondary stream is larger than that of the primary stream, and both streams are mixed immediately in the burner vicinity.

7.3.2 Influence of oxidiser O$_2$ concentration in the secondary stream

Returning to Figure 7.2, one can observe that for O$_2$/CO$_2$ combustion at an oxygen concentration of 18 vol%, the conversion to NO$_x$ is in the order of 25%. With increasing O$_2$ content, there is a slight trend to reduced conversion down to 16% at 27 vol%. The overall conversion of fuel N to NO$_x$ is in the order of 15–30%.

In O$_2$/RFG combustion, NO$_x$ emissions are between 50 and 120 mg/MJ between an O$_2$ concentration of 18 and 21 vol%, thus resulting in fuel-N conversion of about 10%.

Table 7.2 shows NO$_x$ concentrations in the recirculation line measured at a point located after the addition of oxygen to form an O$_2$/RFG mixture. Here the fraction of NO$_2$ is below 3% of total NO$_x$. Hence, the increase of the NO$_x$ concentration in the recirculation line is proportional to the increase inside the furnace. Thus, NO is neither produced nor reduced within the recirculation line.

7.4 Influence of primary stream momentum on NO$_x$ emissions

The transition from flame Type 2 to Type 3 (described in Smoot and Pratt [126]) is characterised by deeper penetration of the primary stream jet into the IRZ, thus affecting

Table 7.2 NO concentration in the recirculated flue gas (air ratio $\lambda = 1.2$, burner air ratio $\lambda_b \approx 0.6$, recycle ratio RR ≈ 0.6).

O$_2$ Conc. in Oxidiser (Vol%)	Oxidiser			Furnace End
	NO (ppm)	NO$_2$ (ppm)	NO$_x$ (mg/MJ)	NO$_x$ (mg/MJ)
18	180	0	108	201
19	203	2	117	215
20	221	3	121	223
21	324	9	173	315
21	337	11	179	314

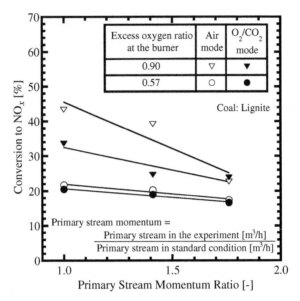

Figure 7.3 Conversion of fuel N as a function of primary stream momentum ratio.
Source: Reprinted from Energy & Fuels, Vol. 26, P. Michitaka Ikeda, Dobrin Toporov, Dominik Christ, Hannes Stadler, Malte Förster, and Reinhold Kneer, Trends in NO_x Emissions during Pulverized Fuel Oxy-fuel Combustion, 3141–3149, (2012), with permission from ACS Publications.

significantly the conditions of particle ignition and NO_x formation. Therefore, the effect of the primary stream momentum on NO_x emissions in PF oxy-firing was investigated experimentally. The primary steam momentum ratio, PM, is defined as follows:

$$PM = PM_{exp}/PM_{std} \qquad (7.3)$$

where PM_{exp} and PM_{std} are the flow rates (in m^3/h) of the primary stream in the experiment and in standard conditions, respectively. Two combustion modes, namely, air and O_2/CO_2, were considered. Figure 7.3 shows the influence of the volume of primary steam upon the conversion to NO_x in the case of two excess oxygen ratios at the burner. The obtained results show that a decrease of primary stream momentum leads to an increase of NO_x conversion for all combustion modes. This effect is more pronounced for the higher local $\lambda = 0.9$ than for $\lambda = 0.57$.

This behaviour can be explained as follows: At lower primary stream momentum, the inertial force in axial direction becomes weak, and PF diffuses easily in radial direction. Therefore, the oxygen concentration is high (secondary stream) and the fuel-N oxidises to NO. Additionally, in air mode, a large amount of thermal NO_x is formed. An increase of the primary stream momentum leads, as expected, to lower NO_x emissions. However, stronger primary stream momentum can lead to lower combustion efficiency due to the fact that the coal particle residence time in the flame becomes shorter. In the case of combustion tests, low combustion efficiency was not indicated. It is known that the combustion efficiency decreases linearly as the fuel ratio, which is the weight ratio of fixed carbon to volatile matter, increases. The fuel ratio of lignite coal, burned here, is much lower than that of bituminous coal. Therefore, even if the combustion condition is not optimised, the combustion efficiency of lignite is kept extremely high and only NO_x is mainly changed. For all experiments, the CO levels were below 50 ppm. In any case, it can be recommended that an adjustment of the primary stream momentum should consider both the NO_x emissions and flame stability and, respectively, carbon burnout.

7.5 Influence of the burner inlet temperature on NO_x emissions

Figure 7.4 shows the influence of the stream temperature from the burner on the conversion to NO_x. The conversion of NO_x slightly increases with the increase in the inlet temperature of the secondary stream. This increase tends to be higher in air mode than in O_2/CO_2 mode. It is considered that the formation of thermal NO_x increases because the flame temperature increases at higher stream temperature. In comparison to the influence of the oxygen concentration in the secondary stream upon the conversion to NO_x, it was found that conversion to NO_x at the higher stream temperature is not reduced. Both the higher oxygen concentration and higher temperature in the secondary stream move the ignition point closer to the burner inlet, and the NO reduction region

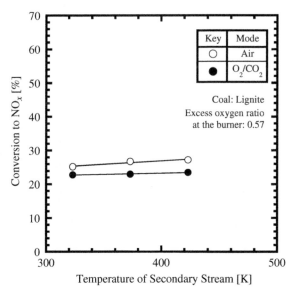

Figure 7.4 Conversion of fuel N as a function of oxidiser inlet temperature.
Source: Reprinted from Energy & Fuels, Vol. 26, P. Michitaka Ikeda, Dobrin Toporov, Dominik Christ, Hannes Stadler, Malte Förster, and Reinhold Kneer, Trends in NO_x Emissions during Pulverized Fuel Oxy-fuel Combustion, 3141–3149, (2012), with permission from ACS Publications.

inside the flame expands. However, the higher stream temperature also causes higher flame temperature. Then the amount of fuel-NO formation increases. Therefore, it is considered that the conversion to NO_x does not decrease with an increasing stream temperature, despite the improvement of the ignition conditions.

7.6   Influence of burner secondary stream momentum on NO_x emissions

This section discusses the influence of the volume of secondary stream on the NO_x emission characteristics. The secondary stream volume is presented as momentum ratio SM, which is defined in analogy with the primary stream momentum ratio as follows:

$$SM = SM_{exp}/SM_{std} \qquad (7.4)$$

where SM_{exp} and SM_{std} are the flow rate in m^3/h of secondary stream in the experiment and in standard conditions, respectively. Two levels of oxygen excess ratio at the burner are considered for all three combustion modes.

Figure 7.5 shows the influence of the volume of secondary stream on the conversion to NO_x. Higher secondary stream momentum leads to higher NO_x emissions in both air and O_2/CO_2 modes. This is because the pulverised coal particles are easy to diffuse near the combustion chamber, owing to the high volume of secondary stream with

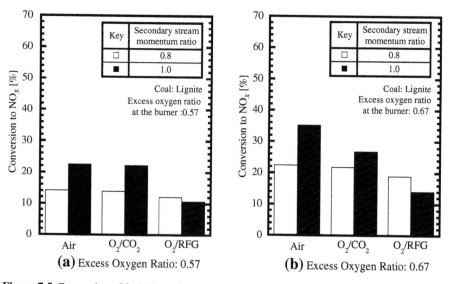

Figure 7.5 Conversion of fuel N as a function of burner secondary stream momentum.
Source: Reprinted from Energy & Fuels, Vol. 26, P. Michitaka Ikeda, Dobrin Toporov, Dominik Christ, Hannes Stadler, Malte Förster, and Reinhold Kneer, Trends in NO_x Emissions during Pulverized Fuel Oxy-fuel Combustion, 3141–3149, (2012), with permission from ACS Publications.

swirl flow, and the nitrogen in coal is oxidised to NO$_x$ at the outer side of the furnace, the O$_2$ concentration of which is higher than that at the inner side. On the other hand, it was found that NO$_x$ emissions in the case of the O$_2$/RFG mode become low as the momentum of secondary stream increases. In this case the high secondary stream momentum also leads to the diffusion of pulverised coal particles and causes a higher NO$_x$ emission near the burner, the same as in air and O$_2$/CO$_2$ modes. However, the secondary (recycled) stream contains NO$_x$. An increase of the SM leads to an increase of the recycled NO$_x$. In the flame, NO$_x$ is in the secondary stream and is able to be destructed by the reduction gas in volatile matter. Therefore, the NO$_x$ emissions can be reduced by increasing the volume of the secondary stream.

From the investigation of the influence of the secondary stream upon the NO$_x$ emission, it was clarified that both the higher O$_2$ concentration in the secondary stream in the range for a given excess oxygen ratio and the higher volume of secondary stream lead to lower NO$_x$ emissions in PF oxycombustion.

7.7 Discussion

7.7.1 Thermal NO

For the swirl flame, the fraction of NO$_x$ emissions that is produced via the thermal NO path can be estimated to 25–30% by evaluating the difference between air and O$_2$/CO$_2$ mode in Figure 7.1. Due to the absence of atmospheric nitrogen in oxyfuel firing, all NO that is produced in this operation mode can be considered fuel NO. However, since the increased CO$_2$ concentration influences the combustion process and since the flame temperature in oxyfuel operation differs from that in air due to the higher molar heat capacity of CO$_2$, the actual amount of fuel-N-to-fuel-NO conversion probably also differs between air and oxyfuel operation so that the difference between NO$_x$ emissions in air and in O$_2$/CO$_2$ combustion gives only a rough estimation of thermal NO.

The fraction considered thermal NO is slightly higher compared to data from standard swirl burners [221,222]. The burner is designed to induce a strong recirculation of hot reaction products to the burner quarl in order to stabilise the flame in oxyfuel mode. This intense recirculation causes the increased formation of thermal NO in air mode. At burner air ratios below $\lambda_b = 0.6$, no thermal NO seems to be produced. This can be attributed to the reduction of the oxygen concentration in the burner proximity at low burner air ratios.

With respect to O$_2$ concentration in the oxidiser, the NO$_x$ emissions follow the same trend as experimental data published earlier [223]. For O$_2$/CO$_2$ mode, NO$_x$ emissions can be considered constant or slightly decreasing with increasing oxidiser oxygen concentration. This is contrary to the trend found for O$_2$/RFG mode and to investigations at higher oxidiser oxygen concentrations by Croiset and Thambimuthu [4] and Williams et al. [224], as well as Hjaertstam et al. [225], who showed an increase of NO$_x$ emissions with increasing O$_2$ content in the oxidiser. However, following the argument by Liu and Okazaki [109] and Normann [106], this stagnation or decrease of NO$_x$ emissions could be interpreted as an indicator of improved thermal NO reduction caused by an increase of the adiabatic flame temperature with increasing O$_2$ content in the oxidiser.

7.7.2   Fuel NO

Several mechanisms affect fuel NO emissions under oxyfuel conditions, partly in opposite directions. Fuel N conversion under oxyfuel conditions is reduced compared to the air mode, which is to some extent due to a reduced oxidation of HCN as reported by Giménez-López et al. [226].

However, the experiments indicated an increase in fuel NO production in O_2/CO_2 mode. This can be explained by higher char NO emissions, which could partly compensate for the reduced oxidation of HCN. This effect was already reported by Spinti and Pershing [227], who found a slightly negative dependence of char-N-to-NO conversion with respect to flame temperature.

Moreover, a high concentration of CO_2 enhances the formation of NO under reducing conditions while it inhibits NO formation under stoichiometric and lean conditions, as reported by Mendiara and Glarborg [146]. These findings could give another explanation for the trend of slightly increased NO_x emissions with reduced oxidiser oxygen concentration shown in Figure 7.2.

7.7.3   In-flame NO_x reduction

The reburning effect due to the recirculation of flue gas back to the flame plays an essential role in NO_x emissions during oxycoal combustion. As the burner oxygen ratio increases, the recycle ratio also increases, which correlates to higher NO_x reduction rates. This effect has already been shown by Hu et al. [228], who also reported that the reduction efficiency of recycled NO_x is improved with decreasing burner oxygen ratio. On the contrary, Dhungel et al. [229] reported no significant difference in dependence of NO_x emissions on burner oxygen ratio between air mode and an O_2/RFG mode with 27 vol% O_2 in the oxidiser.

Commonly, the main cause for the reduction of NO_x emissions in oxyfuel combustion compared to air is attributed to homogeneous reduction of recycled NO_x; see, for example, Okazaki and Ando [107]. However, especially from Table 7.2 it becomes obvious that a reduction of all recycled NO_x would not be sufficient to explain the low NO_x emissions in O_2/RFG mode compared to O_2/CO_2 mode. Thus, next to a high reduction ratio of recycled NO_x, the fuel N conversion ratio is also reduced in O_2/RFG compared to O_2/CO_2. The biggest difference between the two modes is the increased concentration of water vapour in wet O_2/RFG due to the recirculation of the combustion products (between 13 vol% and 14 vol%). Apparently H_2O inhibits the oxidation of intermediates to form NO or the NO reduction with the intermediates is substantially improved. Schäfer and Bonn [230] have shown that the addition of water vapour significantly enhances the hydrolysis of HCN to NH_3. Another effect that reduces NO_x emissions in either oxyfuel mode is that due to the increase of the CO_2 concentration, the CO concentration also increases, which enhances char-catalysed reduction of NO_x, as reported by Okazaki and Ando [107].

At an oxidiser oxygen concentration of 18 vol% the reduction of the fuel N conversion ratio obtained in O_2/RFG mode is more than 30 vol% compared to those obtained in O_2/CO_2 mode, whereas NO_x emissions at an oxidiser oxygen concentration of 21 vol% could as well be explained without a reduction of the fuel N conversion ratio.

An increased oxygen concentration apparently compensates the positive effect of the increased water concentration. One possible explanation is the improved oxidation of intermediate species due to the higher oxygen concentration; a second explanation is the increased temperature at higher oxygen concentrations, which enhances NO$_x$ formation.

From these results, a method of predicting NO$_x$ emissions based on the mathematical model that is calculated by multiplex analysis has been applied for oxyfuel conditions and presented by Ikeda *et al.* [231]. Thus, the trends observed for the formation and destruction of NO$_x$ can be translated to industrial-scale facilities.

7.8   Summary of chapter 7

Experimental investigations on NO$_x$ emissions from the scaled-up burner have been carried out. The main results of these experiments can be summarised as follows:

- Increasing of burner excess oxygen ratio leads to an increase of NO$_x$ formation for all three combustion modes. NO$_x$ emissions in O$_2$/CO$_2$ mode are about 20% lower than in air mode. This causes high temperatures in the burner vicinity to form large amounts of thermal NO$_x$ in air mode. On the other hand, NO$_x$ emissions in O$_2$/RFG mode are strongly reduced, by approximately 50%. This is due to the fact that NO$_x$ contained in RFG is supplied back by secondary stream to the flame and thus destructed by reduction gas in volatile matter.
- A decrease in the volume of the primary stream leads to not only improved combustion but also to an increase in NO$_x$ emissions for all combustion modes.
- An increase in the oxygen concentration in the secondary streams does not affect the global NO$_x$ emission. In this case, the ignition point closes to the burner, and NO$_x$ formation increases in the near burner region. In parallel, the NO$_x$ reduction region inside the flame expands owing to early ignition, thereby increasing the NO$_x$ reduction ratio.
- NO$_x$ emissions increase slightly with the increase in the inlet temperature of the secondary stream. This increase is higher in air mode than in O$_2$/CO$_2$ mode. It is considered that the formation of thermal NO$_x$ increases because the flame temperature increases at the higher stream temperature.
- A higher volume of secondary stream leads to higher NO$_x$ emissions in both air and O$_2$/CO$_2$ modes. In the case of the O$_2$/RFG mode, however, it is found that NO$_x$ emissions decrease with increasing volume of the secondary stream.

This work demonstrates that wet recycling is preferable in terms of NO$_x$ emissions. However, further investigations into the influence of water vapour on NO$_x$ mechanisms during oxyfuel coal combustion are needed.

Although the NO values obtained in a pilot scale are not representative of a large-scale boiler, the trends for NO$_x$ can be translated to industrial-scale facilities.

8 Summary

The study conducted deals with interdisciplinary problems in the field of fluid mechanics, organic chemistry, thermodynamics, numerical methods in applied mathematics, and informatics.

The investigated problem of "combustion of pulverised fuel in a mixture of oxygen and recycled flue gas" is per se with significant degree of complexity due to the complex nature of the processes involved.

Detailed theoretical study of coal particle combustion processes is carried out. The basic mechanisms describing the release and combustion of volatiles, coal particle ignition, coal char oxidation and gasification, the formation of pollutants, and the radiative emittance from coal particles and from combustion gases are reviewed. The factors influencing these mechanisms and the specific effects appearing due to the high content of CO_2 during oxyfuel combustion are identified and analysed.

The flame stability and the related design of the PF burner, the formation of pollutants and their control, and the changes in the heat transfer are considered the main challenges for the engineers who have to design the process of PF combustion in a mixture of oxygen and recycled flue gas.

For this purpose, a CFD-based model for oxyfuel pulverised coal combustion was developed. Basic models for turbulence, turbulence-chemical reactions and interactions, coal particle pyrolysis, and coal char oxidation and gasification, as well as gas emissivity, are reviewed and analysed.

Validation against experimental data obtained for cold flows and for oxycoal flames in swirl burners showed that $k - \epsilon$ and Reynolds stress turbulence models can correctly predict the main flow pattern; however, the peak values of the backflow axial velocity component at the burner axis and of the tangential velocity components inside the IRZ were under-predicted. Therefore, it is expected that future LES-based calculations of oxycoal swirl flames would offer improvements over $k - \epsilon$ and RSM predictions.

Three heterogeneous reactions have been recognised to play the major role in char burnout: char O_2, char CO_2, and char steam. Therefore, detailed validations of char combustion submodels were performed, applying apparent, intrinsic, and Langmuir-based model approaches. The models were validated against experimental data obtained for different coal qualities upon different reactor temperatures and partial pressures of the bulk, thus demonstrating their ability to predict the reaction rate in a wide range of operating conditions. External routines for all these models were developed and integrated into the global CFD code.

Experimental and numerical investigations were carried out on oxyfuel methane combustion with the aim of assessing the importance of the chemical effects due to

Combustion of Pulverised Coal in a Mixture of Oxygen and Recycled Flue Gas. http://dx.doi.org/10.1016/B978-0-08-099998-2.00008-4

high CO_2 concentrations on homogeneous combustion rates. Experiments have been performed in a 25 kW furnace for flameless combustion, which provides the possibilities to achieve stable combustion of methane within a wide range of oxygen concentrations in the CO_2/O_2 mixture at constant reactor temperature. Four different oxidiser mixtures (CO_2/O_2 and N_2/O_2, both with 21 vol% and 18 vol% O_2) have been studied by detailed in-furnace measurements for flue gas compositions and temperature. In the case of combustion in N_2/O_2 atmosphere, the CO profiles obtained for different O_2 concentrations overlap, thus demonstrating that changing the O_2 concentration did not affect combustion rates with keeping the temperature constant. In the case of combustion in CO_2/O_2 atmosphere, the obtained CO concentrations were much higher than those in N_2/O_2 atmosphere. In contrast to N_2/O_2, the O_2 concentrations had a significant impact on the production and consumption rates of CO in oxyfuel combustion. The results demonstrated that by elimination of the influence of molar heat capacity by keeping constant reactor temperature, CO_2 dissociation by keeping reactor temperature below 900 °C, and thermal radiation by achieving flameless combustion in a small-scale furnace, it can be estimated that the high CO_2 leads to (1) an increase in CO production rates and (2) slower consumption rates of CO. An increase of O_2 in oxyfuel led to a reduction of this impact, however, further investigations on the exact mechanism are necessary.

Numerical predictions, based on RANS models and EDC, coupled with existing global kinetic mechanisms, can provide adequate simulation of methane oxidation in O_2/CO_2 atmosphere. However, new detailed mechanisms that consider the chemical effects of CO_2 on the CO reaction rate, are needed.

Further more, the performance of two different PF burner designs, single central orifice type (SCO) and single annular orifice type (SAO), under oxycoal conditions has been examined in a downfired oxycoal pilot plant.

A swirl oxycoal flame was experimentally stabilised at the burner quarl by (1) increasing O_2 concentration above 34 vol% without changes to the air-firing burner design, and (2) by modifications of the burner geometry thus changing its aerodynamics.

A novel fluid dynamics-based approach for oxycoal swirl flame stabilisation, based on a thorough understanding of the underlying interactions between burner aerodynamics and reaction kinetics, is introduced. Measures for stabilisation of an oxycoal swirl flame with O_2 contents equal to and lower than those in air are derived regarding compensation of the effects of higher molar heat capacity, heat consumption, and volume changes due to gasification reactions.

CFD-based burner design applying the measures for stabilisation of an oxycoal swirl flame led to the realisation of a stable flame at 23 vol% O_2 concentrations for SCO burners and at 19 vol% for SAO burners. The SAO burner is scaled up from 40 kW to 100 kW based on the constant residence time-scaling approach; after that it is successfully tested in the pilot plant, thus demonstrating the viability of the scaling approach and of the measures derived.

Visual inspection and global measurements confirmed flame stabilisation at the burner quarl under oxycoal conditions with 21 vol% O_2 in the O_2/CO_2 mixture (dry recycle). Furthermore, the burner was successfully tested in both oxycoal conditions using RFG (wet recycling) and in conventional air combustion. The tests demonstrated the burner's ability to form a stable swirl flame and to provide good burnout in all three combustion modes, namely, in air, in O_2/CO_2, and in O_2/RFG.

The design of an industrial-scale oxy-firing burner in the range of several MWth, however, should be based on a modification of a conventional air-firing utility-scale burner according to the measures for oxycoal swirl flame stabilisation.

Finally, in order to investigate the influence of different burner operation parameters and of the operation mode on NO_x emissions during PF oxycoal combustion, comparative experiments have been carried out with the scaled-up burner described in Section 6.4. Three combustion modes, namely, (1) air; (2) O_2/CO_2, and (3) O_2/RFG have been considered. The impact of five different burner operational parameters has been studied. The main results demonstrated that:

- Increasing of burner excess oxygen ratio leads to an increase of NO_x formation for all three combustion modes. NO_x emissions in O_2/CO_2 mode are about 20% lower than in air mode. This causes high temperatures in the burner vicinity to form large amounts of thermal NO_x in air mode. On the other hand, NO_x emissions in O_2/RFG mode are strongly reduced by approximately 50%. This is due to the fact that NO_x contained in RFG is supplied back by secondary stream to the flame and thus it is destructed by reduction gas in volatile matter.
- A decrease in the volume of primary stream leads to not only improved combustion but also to an increase of NO_x emissions for all combustion modes.
- An increase in the oxygen concentration in the secondary streams does not affect the global NO_x emission. In this case the ignition point closes to the burner and NO_x formation increases in the near burner region. In parallel, the NO_x reduction region inside the flame expands, owing to early ignition, thereby increasing the NO_x reduction ratio.
- NO_x emissions increase slightly with the increase in the inlet temperature of the secondary stream. This increase is higher in air mode than in the O_2/CO_2 mode. It is considered that the formation of thermal NO_x increases because the flame temperature increases at the higher stream temperature.
- A higher volume of secondary stream leads to higher NO_x emissions in both the air and O_2/CO_2 modes. In the case of the O_2/RFG mode, however, it is found that NO_x emissions decrease with the increasing volume of secondary stream.

Based on these results, it can be concluded that wet recycling is preferable in terms of NO_x emissions. However, further investigations on the influence of water vapour on NO_x mechanisms during oxyfuel coal combustion are needed.

9 Outlook

The main motivation for this study was to understand the basic underlying mechanisms that define the oxycoal combustion and, based on this knowledge, to develop a technology for reliable combustion of pulverised fuel in oxy-firing furnaces. Therefore, the main results obtained and reported here can be used as guidelines for design and operation of large, utility-scale burners, thus ensuring stable, controlled, and safe combustion of PF in both air and oxy-firing modes. The combustion technology developed and described here is seen to pave the way to an industrial-scale implementation of the oxyfuel CCS option when PF is fired.

This development should go together with further fundamental research on (1) combustion chemistry in CO_2-rich atmosphere; (2) materials and related corrosion problematics; and (3) efficient production of oxygen. Furthermore, the realisation of pilot and demonstration projects will bring valuable information needed to understand the effects of a CO_2-rich atmosphere on the PF oxy combustion and on the heat transfer in real (industrial) scales and important know-how for the potential scale-up and construction of a CO_2-emission-free coal-fired power plant.

Thus the oxyfuel technology can become a realistic option for bringing a carbon-based economy to carbon-constrained energy production.

The knowledge reported here can be applied not only in the power generation industry but also in the cement and steel production industries because they use coal in their processes and thus are amongst the largest CO_2 emitters.

Finally, the reader should keep in mind that the application of CCS technologies does not solve the global problem of clean energy production. These technologies are viewed as "bridge technologies that will allow the use of still abundant and cheap fuel, such as coal, in an environmentally friendly way until new technologies, based on fundamentally new processes, will be developed and widely applied.

Combustion of Pulverised Coal in a Mixture of Oxygen and Recycled Flue Gas. http://dx.doi.org/10.1016/B978-0-08-099998-2.00009-6

Bibliography

[1] T.F. Wall, Combustion processes for carbon capture, Proceedings of the Combustion Institute 31 (2007) 31–47.

[2] T. Wall, Y. Liu, Ch. Spero, L. Elliott, S. Khare, R. Rathnam, F. Zeenathal, B. Moghtaderi, B. Buhre, Ch. Sheng, R. Gupta, T. Yamada, K. Makino, J. Yu, An overview on oxy-fuel coal combustion – state of the art research and technology development, Chemical Engineering Research and Design 87 (8) (2009) 1003–1016, http://dx.doi.org/10.1016/j.cherd.2009.02.005.

[3] K. Andersson, R.T. Johansson, S. Hjärtstam, F. Johnsson, B. Leckner, Radiation intensity of lignite-fired oxy-fuel flames, Experimental Thermal and Fluid Science 33 (2008) 67–76, http://dx.doi.org/10.1016/j.expthermflusci.2008.07.010.

[4] E. Croiset, K.V. Thambimuthu, NO_x and SO_2 emissions from O_2/CO_2 recycle coal combustion, Fuel 80 (2001) 2117–2121.

[5] M.B. Toftegaard, J. Brix, P. Jensen, P. Glarborg, A. Jensen, Oxy-Fuel Combustion of Solid Fuels, Progress in Energy and Combustion Science 36 (5) (2010) 581–625, http://dx.doi.org/10.1016/j.pecs.2010.02.001.

[6] World Energy Outlook 2010, Internal Energy Agency, 2010.

[7] Reserven, Ressourcen und Verfügbarkeit von Energierohstoffen 2010, Kurzstudie, Bundesanstalt für Geowissenschaften und Rohstoffe (BGR), Hanover, Deutschland, 2010.

[8] BP Statistical Review of World Energy, June 2010.

[9] B. Bruck, Konstruktion und Inbetriebnahme einer Laboranlage zur Torrefikation von Biomasse, RWTH Aachen University, Institute of Heat and Mass Transfer, Diplomarbeit, 2010.

[10] J.M. Beer, Combustion technology developments in power generation in response to environmental challenges, Progress in Energy and Combustion Science 26 (2000) 301–327.

[11] Mikko Hupa, Fluidised bed combustion of biomass waste-derived fuels – curent status and chalenges, in: WTERT 2005 Fall Meeting at Columbia University, October 20, 2005.

[12] The WTA Technology, An Advanced Method of Processing and Drying Lignite, RWE Power, January 2008.

[13] N. Sarunac, M. Ness, Ch. Bullinger, One year of operating experience with a prototype fluidised bed coal dryer at coal creek generating station, in: Proceedings of the Third Inernational Conference on Clean Coal Technologies for our Future, May 15–17, 2007, Cagliari, Sardinia, Italy, 2007.

[14] D. Toporov, Joao Azevedo, A CFD based numerical investigation of the coal quality impact on the combustion process, International Journal on Energy for a Clean, Environment 7 (1) (2006) 1–16.

[15] L.D. Smoot, Fundamentals of Coal Combustion, Elsevier, 1993.

[16] World Energy Outlook 2006, Internal Energy Agency, 2006.

[17] R. Srivastava, R. Hall, S. Khan, K. Culligab, B. Lani, Nitrogen oxides emissions control options for coal-fired electric utility boilers, Journal of the Air and Waste Management Association 55 (2005) 1367–1388.

[18] E. Pacyna, J. Pacyna, F. Steenhuisen, S. Wilson, Global anthropogenic mercury emission inventory for 2000, Atmospheric Environment 40 (22) (2006) 4048–4063.

Combustion of Pulverised Coal in a Mixture of Oxygen and Recycled Flue Gas. http://dx.doi.org/10.1016/B978-0-08-099998-2.00018-7

[19] J. Katzer (Hrsg.), The Future of Coal. Options for a Carbon-Constrained World, Massachusetts Institute of Technology, 2007.

[20] P. Tans, Trends in Atmospheric Carbon Dioxide, NOAA/ESRL, Trends CO_2, March 2011. <www.esrl.noaa.gov/gmd/ccgg/trends/>.

[21] E. Yantovsky, J. Gorski, M. Shokotov, Zero Emissions Power Cycles, first ed., Taylor & Francis, 2009.

[22] D. Adams, J. Davison, Capturing CO_2/Report, International Energy Agency Greenhouse Gas R&D Programme, Forschungsbericht, 2007.

[23] J. Davison, Performance and costs of power plants with capture and storage of CO_2, Energy 32 (2007) 1163–1176.

[24] J. Tranier, N. Perrin, A. Darde, ASU and CO_2 CPU for oxy-combustion, in: International Network for Oxy-Combustion with CO_2 Capture, 3rd Workshop, Yokohama, Japan, 2008.

[25] Ph. Armstrong, T. Foster, D. Bennett, V. Stein, ITM oxygen: scaling up a low-cost oxygen supply technology, in: Gasification Technologies Conference, Washington, DC, October 2006, pp. 1–4.

[26] R. Hassa, Carbon capture and storage a technology for the coal fired power plant of the future, International Journal for Electricity and Heat Generation, VGB PowerTech 12 (2008) 39–41.

[27] R. Kneer, D. Toporov, M. Förster, D. Christ, Ch. Broeckmann, E. Pfaff, M. Zwick, S. Engels, M. Modigell, OXYCOAL-AC: towards an integrated coal-fired power plant process with ion transport membrane-based oxygen supply, Energy and Environmental Science 3 (2) (2010) 198–207, http://dx.doi.org/10.1039/B908501G.

[28] H. Stadler, F. Beggel, M. Habermehl, B. Persigehl, R. Kneer, M. Modigell, P. Jeschke, Oxyfuel coal combustion by efficient integration of oxygen transport membrane, International Journal of Greenhouse Gas Control (2010), http://dx.doi.org/10.1016/j.ijggc.2010.03.004.

[29] F. Beggel, S. Engels, M. Modigell, N. Nauels, Oxyfuel combustion by means of high temperature membranes for air separation, in: Clean Coal Technologies, Dresden, May 2009.

[30] A. Kather, G. Scheffknecht, The oxycoal process with cryogenic oxygen supply, Naturwissenschaften 96 (2009) 993–1010, http://dx.doi.org/10.1007/s00114-009-0557-2.

[31] D. Toporov, M. Förster, R. Kneer, Combustion of pulverized fuel under oxycoal conditions at low oxygen concentrations, in: Third International Conference on Clean Coal Technologies for our Future, Cagliari, Sardinia, Italy, May 15–17, 2007.

[32] F. Kluger, P. Mönckert, B. Krohmer, G. Stamatelopoulos, J. Jacoby, U. Burchhardt, Oxyfuel pulverised coal steam generator development 30 MWth Pilot steam generator commissioning and testing, in: First IEAGHG International Oxyfuel Combustion Conference, Cottbus, Germnay, 2009.

[33] H. Martin (Hrsg.), VDI – Waermeatlas: Berechnungsblaetter fuer den Waermeuebergang, Springer-Verlag, Berlin, 2002.

[34] S. Saxena, Devolatilisation and combustion characteristics of coal particles, Progress in Energy and Combustion Science 16 (1990) 55–94.

[35] K. Smith, L. Smoot, T. Fletcher, R. Pugmire, The Structure and Reaction Processes of Coal, Plenum Press, New York, 1994.

[36] P. Solomon, T. Fletcher, Impact of coal pyrolysis on combustion, in: 25th Symposium (International) on Combustion, The Combustion Institute, 1994, pp. 463–474.

[37] T. Fletcher, D. Hardesty, Compilation of Sandia Coal Devolatilisation Data/Sandia Report No.SAND92-8209, Combustion Research Facility, Sandia National Laboratories, Livermore, Forschungsbericht, US, 1992.

[38] T. Wall, G. Liu, H. Wu, D. Roberts, K. Benfell, S. Gupta, J. Lucas, D. Harris, The effects of pressure on coal reactions during pulverised coal combustion and gasification, Progress in Energy and Combustion Science 28 (2002) 405–433.

[39] P. Solomon, D. Hamblen, R. Carangelo, J. Krause, Coal thermal decomposition in an entrained flow reactor: experiments and theory, in: 19th Symposium (International) on Combustion, The Combustion Institute, 1982, pp. 1139–1149.

[40] D. Genetti, T.H. Fletcher, R.J. Pugmire, Development and application of a correlation of 13C NMR chemical structural analyses of coal based on elemental composition and volatile matter content, Energy and Fuels 13 (1999) 60–68.

[41] M. Suuberg, W. Peters, J. Howard, Product compositions and formation kinetics in rapid pyrolysis of pulverised coal – implications for combustion, in: 17th Symposium (Intenrational) on Combustion, The Combustion Institute, Pittsburgh, 1979, pp. 117–130.

[42] P. Solomon, M. Serio, R. Carangelo, R. Bassilakis, Analysis of the argonne premium coal samples by thermogravimetric Fourier transform infrared spectroscopy, Energy and Fuels 4 (1990) 319–333.

[43] A. Fedorov, T. Khmel, Y. Gosteev, Theoretical investigation of ignition and detonation of coal-particle gas mixtures, Shock Waves 13 (2004) 453–463.

[44] Anna Ponzio, Sivalingam Senthoorselvan, Weihong Yang, Wlodzmierz Blasiak, Ola Eriksson, Ignition of single particles in high-temperature oxidiziers with various oxygen concentrations, Fuel 87 (2008) 974–987.

[45] D. Parkins, Ignition of coal and char particles: effects of pore structure and process conditions (Dissertation), Rice University, Houstan, Texas, 1998.

[46] M. Baum, P. Street, Predicting the combustion behaviour of coal partilces, Combustion Science and Technology 3 (1971) 231–243.

[47] V. Gururajan, T. Wall, R. Gupta, J. Truelove, Mechanisms for ignition of pulverised coal particles, Combustion and Flame 81 (1990) 119–132.

[48] R. Mitchel, R. Kee, P. Glarborg, M. Coltrin, The effect of CO conversion in the boundary layers surrounding pulverised-coal char particles, in: 23rd Symposium (International) on Combustion, The Combustion Institute, 1990, pp. 1169–1176.

[49] L. Duan, C. Zhao, W. Zhou, C. Qu, X. Chen, Investigation on coal pyrolysis in CO_2 atmosphere, Energy and Fuels 23 (2009) 3826–3830.

[50] R. Messenboeck, D. Dugwell, R. Kandiyoti, CO_2 and steam-gasification in a high-pressure wire-mesh reactor: the reactivity of Daw Mill coal and combustion reactivity of its char, Fuel 78 (1999) 781–793.

[51] H. Liu, M. Kaneko, Ch. Luo, Sh. Kato, T. Kojima, Effect of pyrolysis time on the gasification reactivity of char with CO_2 at elevated temperatures, Fuel 83 (2004) 1055–1061.

[52] K. Jamil, J. Hayashi, C. Li, Pyrolysis of a Victorian brown coal and gasification of nascent char in CO_2 atmosphere in a wire-mesh reactor, Fuel 83 (2004) 833–843.

[53] Ch.R. Shaddix, A. Molina, Particle imaging of ignition and devolatilization of pulverized coal during oxy-fuel combustion, Proceedings of the Combustion Institute 32 (2009) 2091–2098.

[54] A. Yamamoto, T. Suda, K. Okazaki, Mechanism of ignition delay in O_2/CO_2 pulverised coal combustion, in: 21th Annual Pittsburgh Coal Conference, Osaka, Japan, 2004.

[55] T. Suda, M. Katsumi, J. Sato, A. Yamamato, K. Okazaki, Effect of carbon dioxide on flame propagation of pulverised coal clouds in CO_2/O_2 combustion, Fuel 86 (2007) 2008–2015.

[56] A. Molina, E. Hecht, C. Shaddix, Ignition of a group of coal particles in oxyfuel combustion with CO_2 recirculation, in: Clearwater Clean Coal Conference, 2009.

[57] N. Laurendeau, Heterogeneous kinetics of coal char gasification and combustion, Progress in Energy and Combustion Science 4 (1978) 221–270.

[58] R. Hurt, Structure, properties, and reactivity of solid fuels, in: 27th Symposium (International) on Combustion, The Combustion Institute, 1998, pp. 2887–2904.

[59] R.H. Essenhigh, Rate equations for the carbon-oxygen reaction: an evaluation of the Langmuir adsoption isotherm at atmospheric pressure, Energ and Fuel 5 (1991) 41–46.

[60] K. Miura, K. Hashimoto, P. Silverston, Factors affecting the reactivity of coal chars during gasification, and indices representing reactivity, Fuel 68 (1989) 1461.

[61] J.D. Blackwood, F.K. McTaggart, Reaction of carbon with atomic gases, Australian Journal of Chemistry 12 (4) (1959) 533–542.

[62] S. Dutta, C. Wen, R. Belt, Reactivity of coal char in carbon dioxide atmosphere, Industrial and Engineering Chemistry Process Design and Development 16 (1) (1977) 20.

[63] N. Russel, J. Gibbins, J. Williamson, Structural ordering in high temperature coal chars and the effect on reactivity, Fuel 78 (1999) 803–807.

[64] T.H. Fletcher, A.R. Kerstein, R.J. Pugmire, M. Solum, D.M. Grant, A Chemical Percolation Model for Devolatilization: Summary/Sandia Technical Report Sand 92–8207, Forschungsbericht, 1992.

[65] L. Smooth, P. Smith, Heterogeneous Char Reaction Processes in Coal Combustion and Gasification, Plenum Press, New York, NY, 1985.

[66] A. Blokh, Heat Transfer in Steam Boilers, Hemisphere, Washington, DC, 1988.

[67] N. Laine, F. Vastola, P. Walker, The importance of active surface area in the carbon-oxygen reaction, Journal of Physical Chemistry 67 (1963) 2030.

[68] R. Mitchell, Variations in the temperatures of coal-char particles during combustion: a consequence of particle-to particle variations in ash-content, in: 23rd Symposium (International) on Combustion, The Combustion Institute, Pittsburgh, PA, 1990, pp. 1297–1304.

[69] E. Hipo, P. Walker, Reactivity of heat-treated coals in carbon dioxide at 900 °C, Fuel 54 (1975) 245.

[70] R. Hurt, J.-K. Sun, M. Lunden, A kinetic model of carbon burnout in pulverized coal combustion, Combustion and Flame 113 (1998) 181–197.

[71] Y. Yang, A. Watkinson, Gasification reactivity of some Western Canadian coals, Fuel 73 (1994) 1786–1791.

[72] T. Alvarez, A. Fuertes, J. Pis, P. Ehrburger, Influence of coal oxidation upon char gasification reactivity, Fuel 74 (1995) 729–735.

[73] S. Kasaoka, Y. Sakata, M. Shimada, Effects of coal carbonisation conditions on rate of steam gasification of char, Fuel 66 (1987) 697–701.

[74] K. Hashimoto, K. Miura, T. Ueda, Correlation of gasification rates of various coals measured by a rapid heating method in a steam atmosphere at relatively low temperatures, Fuel 65 (1986) 1516–1523.

[75] T. Adschiri, T. Shiraha, T. Kojima, T. Furusawa, Prediction of CO_2 gasification rate of char in fluidised bed gasifier, Fuel 65 (1986) 1688–1693.

[76] H. Wu, G. Bryant, K. Benfell, T. Wall, An experimental study on the effect of system pressure on char structure of an Australian bituminous coal, Energy and Fuels 14 (2000) 282–290.

[77] R. Mitchell, O. Madsen in: Proceedings of the 21st Symposium (International) on Combustion, The Combustion Institute, Pittsburgh, PA, 1986, p. 173.

[78] C. Wang, G. Berry, C. Chang, A.W. Wolsky, Combustion of pulverised coal using waste carbon dioxide and oxygen, Combustion and Flame 72 (1988) 301–310.

[79] L. Kurylko, R. Essenhigh in: Proceedings of the 14th Symposium (International) on Combustion, The Combustion Institute, Pittsburgh, PA, 1975, p. 1375.

[80] M.A. Field, Rate of combustion of size-graded fractions of char from a low-rank coal between 1200 K and 2000 K, Combustion and Flame 13 (1969) (2000) 237–252.

[81] M.A. Field, Measurements of the effect of rank on combustion rates of pulverised coal, Combustion and Flame 14 (1970) 237–248.

[82] M. Mulcahy, I. Smith, Kinetics of combustion of pulverised fuel: a review of theory and experiment, Reviews of Pure and Applied Chemistry 19 (1969) 81–108.

[83] Jeffrey J. Murphy, Christopher R. Shaddix, Combustion kinetics of coal chars in oxygen-enriched environments, Combustion and Flame 144 (2006) 710–729.

[84] L. Shufen, S. Ruizheng, Kinetic studies of a lignite char pressurised gasification with CO_2, H_2 and steam, Fuel 73 (1994) 413–416.

[85] S. Tanaka, T. Uemura, K. Ichizaki, K. Nagayoshi, N. Ikenaga, H. Ohme, T. Suzuki, H. Yamashita, M. Ampo, CO_2 gasification of iron-loaded carbons: activation of the iron catalyst with CO, Energy and Fuels 9 (1995) 45–52.

[86] M. Barrio, J.E. Hustad, Chapter 3. CO_2 gasification of birch char and the effect of CO inhibition on the calculation of chemical kinetics, Progress in Thermochemical Biomass Conversion, A.V. Bridgwater (Ed.), Blackwell Science Ltd., 2001.

[87] G. Ballal, A Study of Char Gasification Reactions, Rice University Diplomarbeit, Houston, Texas, 1985.

[88] S. Schiebahn, Numerische Untersuchung des Einflusses der Kohlendioxidvergasung auf den OXYCOAL-Prozess, Institute of Heat and Mass Transfer, RWTH Aachen Univesity, Diplomarbeit, Germany, 2008.

[89] J. Moulijn, F. Kapteijn, Towards a unified theory of reactions of carbon with oxygen-containing molecules, Carbon 33 (1995) 1155–1165.

[90] D. Roberts, D. Harris, Char gasification in mixtures of CO_2 and H_2O: competition and inhibition, Fuel 86 (2007) 2672–2678.

[91] J. Richard, M.A. Majthoub, M. Aho, P. Pirkonen, Separate effects of pressure and some other variables on char combustion under fixed bed conditions, Fuel 73 (1994) 485–491.

[92] C. Monson, G. Germane, A. Blackham, D. Smoot, Char oxidation at elevated presure, Combustion and Flame 100 (1995) 669–683.

[93] D. Roberts, D. Harris, Char gasification with O_2, CO_2, and H_2O: effects of pressure on intrinsic reaction kinetics, Energy and Fuels 14 (2000) 483–489.

[94] Ch. Shaddix, A. Molina, Effects of O_2 and high CO_2 concentrations on pc char burning rates during oxy-fuel combustion, in: Proceedings of the 33rd International Technical Conference on Coal Utilisation and Fuel Sytems, Clearwater, FL, June 1–5, 2008.

[95] T. Wall, Y. Liu, C. Spero, L. Elliott, S. Khare, R. Rathnam, F. Zeenathal, B. Moghtaderi, B. Buhre, C. Sheng, R. Gupta, T. Yamada, K. Makino, J. Yu, An overview on oxyfuel coal combustion - state of the art research and technology development, Chemical Engineering Research and Design 87 (8) (2009) 1003–1016.

[96] C. Shaddix, S. Jimenez, S. Lee, Evaluation of rank effects and gas temperature on coal char burning rates during oxy-fuel combustion, in: Proceedings of the 34th International Technical Conference on Coal Utilisation and Fuel Systems, Clearwater, FL, 2009.

[97] M. Geier, E. Hecht, C. Shaddix, Evaluation of effect of particle size on oxy-fuel combustion of pulverised coal, in: 26th Annual International Pittsburgh Coal Conference, Pittsburgh, PA, September 2009, pp. 20–23.

[98] D. Zhu, F. Egolfopoulos, C. Law in: 22nd Symposium (International) on COmbustion/Combustion Institute Pittsbourgh, 1988, pp. 1537–1545.

[99] Y. Tan, M.A. Douglas, K.V. Thambimuthu, CO_2 capture using oxygen enhanced combustion strategies for natural gas power plants, Fuel 81 (2002) 1007–1016.

[100] K. Andersson, F. Johnsson, Flame and radiation characteristics of gas-fired O_2/CO_2 combustion, Fuel 86 (2007) 656–668.

[101] F. Liu, H. Guo, G. Smallwood, The chemical effect of CO_2 replacement of N_2 in air on the burning velocity of CH_4 and H_2 premixed flames, Combustion and Flame 133 (2003) 495–497.

[102] P. Glarborg, L. Bentzen, Chemical effects of a high CO concentration in oxy-fuel combustion of methane, Energy and Fuels 22 (1) (2008) 291–296, http://dx.doi.org/10.1021/ef7005854.

[103] P. Canning, A. Jones, P. Balmbridge, NO_x control for large coal-fired utility boilers: selection of the most appropriate technology, Australian Coal Review (1999) 35–42.

[104] D. Toporov, P. Bocian, P. Heil, A. Kellermann, H. Stadler, S. Tschunko, M. Förster, R. Kneer, Detailed investigation of a pulverized fuel swirl flame in CO_2/O_2 atmosphere, Combustion and Flame 155 (2008) 605–618, http://dx.doi.org/10.1016/j.combustflame.2008.05.008.

[105] T. Nozaki, Sh. Takano, T. Kiga, K. Omata, N. Kimura, Analysis of the flame formed during oxidation of pulverized coal by an O_2-CO_2 mixture, Energy 22 (2/3) (1997) 199–205.

[106] F. Normann, K. Andersson, B. Leckner, F. Johnsson, Emission control of nitrogen oxides in the oxy-fuel process, Progress in Energy and Combustion Science 35 (2009) 385–397.

[107] K. Okazaki, T. Ando, NO_x-reduction mechanism in coal combustion with recycled CO_2, Energy 22 (1997) 207–215.

[108] Y.Q. Hu, N. Kobayashi, M. Hasatani, Effects of coal properties on recycle-NO_x reduction in coal combustion with O_2/recycled flue gas, Energy Conversion and Management 44 (2003) 2331–2340.

[109] H. Liu, K. Okazaki, Simultaneous easy CO_2 recovery and drastic reduction of SO_x and NO_x in O_2/CO_2 coal combustion with heat recirculation, Fuel 82 (2003) 1427–1436.

[110] H. Liu, R. Zailani, B.M. Gibbs, Comparisons of pulverized coal combustion in air and in mixtures of O_2/CO_2, Fuel 84 (2005) 833–840.

[111] Dong-Chan Park, Stuart J. Day, Peter F. Nelson, Nitrogen release during reaction of coal char with O_2, CO_2 and H_2O, Proceedings of the Combustion Institute 30 (2005) 2169–2175.

[112] C. Shaddix, A. Molina, NO_x formation in laboratory investigations of oxy-coal combustion, in: 2007 Fall Meeting of the Western States Section of the Combustion Institute, Sandia National Laboratories, Livermore, CA, October 16–17, 2007.

[113] F. Wigley, B. Goh, Characterisation of rig deposits from oxy-coal combustion, in: First Oxyfuel Combustion Conference, Cottbus, September 2009.

[114] S. Schnurrer, L. Elliot, T. Wall in: Proceedings GCHT-7, Newcastle, Australia, June 23–25, 2008.

[115] M. Muller in: Proceedings GCHT-7, Newcastle, Australia, June 23–25, 2008.

[116] A. Kellermann, P. Heil, M. Förster, R. Kneer, Hot gas cleaning at $\approx$ °C[800] of recirculated flue gas in a coal fired high temperature membrane based oxyfuel process: OXYCOAL-AC, in: Seventh International Symposium on Gas Cleaning at High Temperatures. Newcastle, Australien, June 2008.

[117] A. Kellermann, M. Habermehl, M. Förster, R. Kneer, Erste Betriebstests einer Oxycoal-Heißgasreinigung 24, Deutscher Flammentag, Bochum, 2009.

[118] H. Hottel, A. Sarofim, Radiative Transfer, McGraw-Hill, New York, 1967.

[119] R. Goulard (Hrsg.), R. Thompson (Hrsg.), Handbook of Infrared Radiation from Combustion Gases, Scientific and Technical Information Office NASA, Washington, DC, 1973.

[120] R. Johansson, Modelling elements in conversion of solid fuels – fixed bed combustion and gaseous radiation (Dissertation), Chalmers University of Technology, 2008.

[121] R. Gupta, S. Khare, T. Wall, K. Eriksson, D. Lundstrom, J. Eriksson, Ch. Spero, Adaptation of gas emissivity models for CFD based radiative transfer in large air-fired and oxy-fired furnaces, in: 31st International Technical Conference on Coal Utilization and Fuel Systems, Clearwater, Florida, May 2006, pp. 21–26.

[122] R. Payne, S.L. Chen, A.M. Wolsky, W.F. Richter, CO_2 recovery via coal combustion in mixtures of oxygen and recycled flue gas, Combustion Science and Technology 67 (1989) 1–16.

[123] J. Erfurth, D. Toporov, M. Förster, R. Kneer, Numerical simulation of a 1200 MWth pulverised fuel oxy-firing furnace, in: Fourth International Conference on Clean Coal Technology, Dresden, May 2009.

[124] J. Erfurth, D. Toporov, M. Förster, R. Kneer, Simulation der Strahlungswärmeübertragung in einem Oxycoal-Großdampferzeuger 24, Deutscher Flammentag, Bochum, 2009, pp. 139–144.

[125] M. Gibson, B. Morgan, Mathematical model of combustion of solid particles in a turbulent stream with recirculation, Journal of the Institute of Fuel 43 (1970) 517–523.

[126] D. Smoot, D. Pratt, Pulverised-Coal Combustion and Gasification, Theory and Applications for Continuous Flow Processes, Plenum Press, 1979.

[127] F. Lockwood, A. Salooja, S. Syed, A Prediction method for coal-fired furnaces, Combustion and Flame 38 (1980) 1–15.

[128] J. Truelove, The modelling of flow and combustion in swirled pulverised-coal burners, in: 20th Symposium (International) on Combustion, The Combustion Institute, 1984, pp. 523–530.

[129] W. Fiveland, W. Oberjohn, D. Cornelius, COMO: A Numerical Model for Predicting Furnace Performance in Axisymmetric Geometries, Volume I – Technical Summary, Report No. DOE/PC/40265-9/The Babcock & Wilcox Company, R&D Division, Alliance RC, Ohio, Forschungsbericht, 1985.

[130] A. Williams, R. Backreedy, R. Habib, J.M. Jones, M. Pourkashanian, Modelling coal combustion: the current position, Fuel 81 (2002) 605–618.

[131] S. Niksa, Pennsylvania S. University (Hrsg.), Coal Combustion Modelling, IEA Coal Research, 1996.

[132] R. Weber, B.M. Visser, A.A.F. Peters, P.P. Breithaupt, Mathematical modelling of swirling flames of pulverised coal: what can combustion engineers expect from modelling? ASME Journal of Fluids Engineering 117 (1995) 289–297.

[133] U. Schnell, Berechnung der Stickoxidemissionen von Kohlenstaubfeuerungen (Dissertation), University of Stutgart, 1991.

[134] J.L.T. Azevedo, Modelacao Fisica e Simulacao Numerica de Sistemas de Queima de Combustiveis Solidos (Dissertation), Universidade Technica de Lisboa, Instituto Superior Tecnico, 1994.

[135] B. Launder, B. Spalding, The numerical computation of turbulent flows, Computer Methods in Applied Mechanics and Engineering 3 (2) (1974) 269–289.

[136] I. Ertesvag, Turbulent Flow and Combustion, Tapir Academic Publisher, Trondheim, Norway, 2000.

[137] M. Gibson, B. Launder, Ground effects on pressure fluctuations in the atmospheric boundary layer, Journal of Fluid Mechanics 86 (1978) 491–511.

[138] M. Gibson, B. Younis, Calculation of swirling jets with a Reynolds stress closure, Physics of Fluids 29 (1986) 38–48.

[139] S. Jakirlic, Reynolds-Spannungs-Modellierung Komplexer Turbulenter Strömungen (Dissertation), Universität Erlangen-Nürnberg, Diplomarbeit, 1997.

[140] B.J. Daly, F.H. Harlow, Transport equations in turbulence, Physics of Fluids 18 (1970) 2634–2649.

[141] Th. Poinsot, D. Veynante, Thierry Poinsot (Hrsg.), Theoretical and Numerical Combustion, Edwards, Inc., 2005.

[142] H. Watanabe, K. Tanno, Y. Baba, R. Kurose, S. Komori, Large-eddy Simulation of Coal Combustion in a Pulverized Coal Combustion Furnace with a Complex Burner, in: K. Hanjalic, Y. Nagano, S. Jakirlic (Eds.), Turbulence, Heat and Mass Transfer, vol. 6, Beggel House, Inc., 2009.

[143] L.G. Loicianskii, Fluid Mechanics, Science, Moscow, 1970 (in Russian).

[144] I. Shepherd, J. Moss, K. Bray, Turbulent transport in a confined premixed flame, in: 19th Symposium (International) on Combustion, The Combustion Institute, 1982, pp. 423–431.

[145] T. Elperin, N. Kleeorin, I. Rogachevskii, Effect of chemical reactions and phase transitions on turbulent transport of particles and gases, Physical Review Letters 80 (1998) 69–72.

[146] Teresa Mendiara, Peter Glarborg, Ammonia chemistry in oxy-fuel combustion of methane, Combustion and Flame 156 (2009) 1937–1949, http://dx.doi.org/10.1016/ j.combustflame.2009.07.006.

[147] T. Mendiara, P. Glarborg, Reburn chemistry in oxy-fuel combustion of methane, Energy and Fuels 23 (2009) 3565–3572.

[148] R. Borghi, Turbulent combustion modelling, Progress in Energy and Combustion Science 14 (1988) 245–292.

[149] P. Libby, F. Williams, Turbulent Reacting Flows, Academic Press, London, 1994.

[150] D. Spalding, Development of the eddy-break-up model of turbulent combustion, in: 16th Symposium (International) on Combustion, The Combustion Institute, 1976, pp. 1657–1663.

[151] B. Magnussen, B. Hjertager, On mathematical modelling of turbulent combustion with special emphasis an soot formation and combustion, in: 16th Symposium (International) on Combustion, 1977, pp. 719–729.

[152] B. Magnussen, The eddy dissipation concept a bridge between science and technology, in: ECCOMAS Thematic Conference on Computational Combustion, Lisbon, Portugal, 2005.

[153] N. Peters, Laminar flamelet concepts in turbulent combustion, in: Proceedings of the Combustion Institute 21 (1986) 1231–1250.

[154] H. Pitsch, N. Peters, A consistent flamelet formulation for non-premixed combustion considering differential diffusion effects, Combustion and Flame 114 (1998) 26–40.

[155] S. Pope, PDF Methods for Turbulent Reacting Flows, Progress in Energy and Combustion Science 11 (1985) 119–192.

[156] S. Pope, Computation of turbulent combustion: progress and chalenges, in: 23rd Symposium (International) on Combustion, The Combustion Institute, 1990, pp. 591–612.

[157] B. Magnussen, B. Hjertager, J. Olsen, D. Bhaduri, Effects of turbulent structure and local concentrations on soot formation and combustion in C_2H_2 diffusion flames, in: 17th Symposium (International) on Combustion, The Combustion Institute, 1978, pp. 1383–1391.

[158] I. Gran, M. Melaaen, B. Magnussen, Numerical simulation of local extinction effects in turbulent combustor flows of methane and air, in: 25th Symposium (International) on Combustion, The Combustion Institute, 1994, pp. 1283–1291.

[159] I. Gran, B. Magnussen, A numerical study of a bluff-body stabilized diffusion flame. Part 2. Influence of combustion modelling and finite-rate chemistry, Combustion Science and Technology 119 (1996) 191–217.

[160] H. Magel, U. Schnell, K. Hein, Simulation of detailed chemistry in a turbulent combustor flow, in: 26th Symposium (International) on Combustion, The Combustion Institute, 1996, pp. 67–74.

[161] D. Toporov, Numerical modelling of turbulent combustion of natural gas in industrial furnaces (Dissertation), Technical University of Sofia, 2000 (in Bulgarian).

[162] D. Toporov, Z. Liu, J. Azevedo, A numerical investigation of the influence of different combustion models on pulverised coal flames simulations, in: Computational Combustion, ECCOMAS Thematic Conference, Lisbon, Portugal, June 21–25, 2005.

[163] I. Ertesvag, B. Magnussen, The eddy dissipation turbulence energy cascade model, Combustion Science and Technology 159 (2000) 213–235.

[164] H.C. Magel, U. Schnell, K.R.G. Hein, Simulation of detailed chemistry in a turbulent combustor flow, in: Proceedings of the Combustion Institute 26 (1996) 67–74.

[165] G. Gouesbet, A. Berlemont, Eulerian and Lagrangian aproaches for predicting the behaviour of discree particles in turbulent flows, Progress in Energy and Combustion Science 25 (1999) 133–159.

[166] J. Shirolkar, C. Coimbra, M. McQuay, Fundamental aspects of modelling turbulent particle dispersion in dilute flows, Progress in Energy and Combustion Science 22 (1996) 363–399.

[167] D.M. Snider, An incompressible three-dimensional multiphase particle-in-cell model for dense particle flows, Journal of Computational Physics 170 (2001) 523–549.

[168] C. Crowe, D. Stock, M. Sharma, The particle-source-in-cell/PSI-CELL/model for gas-droplet flows, ASME Journal of Fluid Engineering 99 (1977) 325–332.

[169] M.J. Andrews, P.J. O'Rourke, The multiphase particle-in-cell (MP-PIC) method for dense particle flows, International Journal on Multiphase Flow 22 (1996) 379.

[170] S. Badzioch, P. Hawksley, Kinetics of thermal decomposition of pulverized coal particles, Industrial and Engineering Chemistry Process Design and Development 9 (4) (1970) 521–530.

[171] P. Goldberg, R. Essenhigh, Coal combustion in a jet-mix stirred reactor, in: 17th Symposium (International) on Combustion, The Combustion Institute, 1979, pp. 145–154.

[172] H. Kobayashi, J.B. Howard, A.F. Sarofim, Coal devolatilization at high temperatures, in: Symposium (International) on Combustion 16 (1977) 411–425.

[173] A. Jamalludin, J. Truelove, T. Wall, Modelling of coal devolatilisation and its effect on combustion calculations, Combustion and Flame 62 (1985) 85–89.

[174] A. Jamalludin, T. Wall, J. Truelove, Modelling of devolatilisation and combustion of pulverised coal under rapid heating conditions, Coal Science and Chemistry (1987) 61–109.

[175] S.K. Ubhayakar, D.B. Stickler, C.W. Rosenberg, R.E. Gannon, Rapid devolatilization of pulverized coal in hot combustion gases, Symposium (International) on Combustion 16 (1977) 427–436.

[176] St. Niksa, FLASHCHAIN theory for rapid coal devolatilisation kinetics. 5. Interpreting rates of devolatilisation for various coal types and operating conditions, Energy and Fuels 8 (1994) 671–679.

[177] P.R. Solomon, D.G. Hamblen, R.M. Carangelo, M.A. Serio, G.V. Deshpande, General model of coal devolatilisation, Energy and Fuels 2 (1988) 405–422.

[178] D.M. Grant, R.J. Pugmire, T.H. Fletcher, A.R. Kerstein, Chemical model of coal devolatilisation using percolation lattice statistics, Energy and Fuels 3 (1989) 175–186.

[179] Th. Fletcher, D. Hardesty, Compilation of SANDIA Coal Devolatilisation Data, Milestone Report No. SAND92-8209, May 1992, Combustion Research Facility, Sandia National Laboratories, Livermore, Forschungsbericht, 1992.

[180] I. Smith, The combustion rates of coal chars: a review, Proceedings of the Combustion Institute 19 (1982) 1045–1065.

[181] K. Smith, L. Smoot, T. Fletcher, Coal characteristics, structure, and reaction rates, in: L.D. Smoot (Ed.), Fundamentals of Coal Combustion, Elsevier, New York, 1993, pp. 131–298 (Chapter 3).

[182] J. Sun, R. Hurt, Mechanisms of extinction and near-extinction in pulverised solid fuels combustion, Proceedings of the Combustion Institute 28 (2000) 2205–2213.

[183] J. Moors, V. Banin, J. Haas, R. Weber, A. Veefkind, Prediction and validation of burnout curves for Gottelborn char using reaction kinetic determined in shock tube experiments, Fuel 78 (1999) 25–29.

[184] R.H. Essenhigh, An integration path for the carbon-oxygen reaction with internal reaction, Symposium (International) on Combustion 22 (1988) 89–96.

[185] J.R. Arthur, Reactions between carbon and oxygen, Transactions of the Faraday Society 47 (1951) 164–178.

[186] L. Tognotti, J. Longwell, A. Sarofim, The products of the high temperature oxidation of a single char particle in an electrodynamic balance, in: Proceedings of the 23th International Symposium on Combustion, 22–27 July, 1990, pp. 1207–1213.

[187] T.F. Smith, Z.F. Shen, J.N. Friedman, Evaluation of coefficients for the weighted sum of gray gases model, Journal of Heat Transfer 104 (1982) 602–608.

[188] G. Hargrave, M. Pourkashannian, A. Williams, The combustion and gasification of coke and coal chars, Proceedings of the Combustion Institute 21 (1986) 221.

[189] R.H. Essenhigh, A. Mescher, Mechanism of carbon combustion: relative influence of adsorption, desorption and boundary layer diffusion as a function of pressure, Combustion and Flame 111 (1997) 350–352.

[190] J.H.J. Moors, V.E. Banin, J.H.P. Haas, R. Weber, A. Veefkind, Prediction and validation of burnout curves for Göttelborn char using reaction kinetics determined in shock tube experiments, Fuel 78 (1) (1999) 25–29.

[191] R. Hurt, R. Mitchell, Unified high-temperature char combustion kinetics for a suite of coals of various rank, in: 24th Symposium (International) on Combustion, The Combustion Institute, 1992, pp. 1243–1250.

[192] M. Costa, Ch. Sheng, D. Toporov, J. Azevedo, Development of Improved Combustion Engineering Models – Application to NO_x-Reduction Processes, Final Report/Contract No. JOF3-CT97-0043, December 1997–November 2001, July III, Forschungsbericht, 2001.

[193] J.P. Liddy, D.C. Newey, T. Wilson, The combustion behaviour of low rank coals in a drop tube furnace, in: J.A. Moulijn et al. (Eds.), International Conference on Coal Science, Elsevier Science Publishers, 1987.

[194] D.H. Ahn, B.M. Gibbs, K.H. Ko, J.J. Kim, Gasification kinetics of an Indonesian sub-bituminous coal-char with CO_2 at elevated pressure, Fuel 80 (2001) 1651–1658.

[195] P. Mann, J. Kent, A computational study of heterogeneous char reactions in a full-scale furnace, Combustion and Flame 99 (1994) 147–156.

[196] A. Mayers, The rate of reduction of carbon dioxide by graphite, Journal of the American Chemical Society 56 (1934) 70–76.

[197] S. Kajitani, N. Suzuki, M. Ashizawa, S. Hara, CO_2 gasification rate analysis of coal char in entrained flow coal gasifier, Fuel 85 (2006) 163–169.

[198] S. Kajitani, S. Hara, H. Matsuda, Gasification rate analysis of coal char with a pressurized drop tube furnace, Fuel 81 (2002) 539–546.

[199] S. Rumpel, Die authotherme Wirbelschichtpyrolise zur Erzeugung heizwertreicher Stuetzbrennstoffe (Dissertation), University Karlsruhe (TH), Diplomarbeit, 2000.

[200] M.S. Dershowitz, High temperature gasification of coal char in carbon dioxide and steam, Massachusetts Institute of Technology, Diplomarbeit, 1979.

[201] F. Lockwood, N. Shah, A new radiation solution method for incorporation in general combustion prediction procedures, in: 18th Symposium (International) on Combustion, The Combustion Institute, 1981, p. 1405.

[202] D.K. Edwards, Molecular gas band radiation, Advances in Heat Transfer 12 (1976) 115–193.

[203] B. Leckner, Spectral and total emissivity of vapor and carbon dioxide, Combustion and Flame 19 (1972) 33–48.

[204] H.C. Hottel, E.S. Cohen, Radiant heat exchange in a gas-filled enclosure: allowance for nonuniformity of gas temperature, AIChE Journal 4 (1958) 3–13.

[205] P. Heil, Investigation of the flameless combustion of methane in an oxyfuel atmosphere (in German) (Dissertation), RWTH Aachen University, Germany, 2010.

[206] Hiroshi Tsuji, Ashwani K. Gupta, Toshiaki Hasegawa, Masashi Katsuki, Ken Kishimoto, Mitsunobu Morita, Ashwani K. Gupta (Hrsg.), David G. Lilley (Hrsg.), High Temperature Air Combustion, CRC Press, 2003.

[207] M. Katsuki, T. Hasegawa, The science and technology of combustion in highly preheated air, in: 27th Symposium (International) on Combustion, 1998, pp. 3135–3146.

[208] J.A. Wünning, J.G. Wünning, Flameless oxidation to reduce thermal NO-formation, Progress in Energy and Combustion Science 23 (1) (1997) 81–94.

[209] Tobias Plessing, Norbert Peters, Joachim G. Wünning, Laseroptical investigation of highly preheated combustion with strong exhaust gas recirculation, in: 27th Symposium (International) on Combustion, 1998, pp. 3197–3204.

[210] R. Weber, S. Orsino, N. Lallemant, A. Verlaan, Combustion of natural gas with high-temperature air and large quantities of flue gas, Proceedings of the Combustion Institute 28 (2000) 1315–1321.

[211] J. Wünning, Flameless oxidation, in: Sixth HiTACG Symposium, Essen, October 17–19, 2005.

[212] D. Toporov, M. Förster, R. Kneer, Burning pulverized coal in CO_2 atmosphere at low oxygen concentrations, Clean Air 8 (2007) 321–338.

[213] J. Erfurth, D. Toporov, M. Förster, R. Kneer, Oxycoal swirl burners: from bench to industrial scale, in: 10th Conference on Energy for a Clean Environment, Lisbon, Portugal, 2009.

[214] P.A. Jensen, P.R. Ereaut, S. Clausen, O. Rathmann, Local measurements of velocity, temperature and gas composition in a pulverised-coal flame, Journal of the Institute of Energy 67 (1994) 37–46.

[215] Malte Förster, Dobrin Toporov, Reinhold Kneer, Verfahren zur Verbrennung von pulverisiertem Brennmaterial, (05 2007), Pat.-Nr. DE 102007021799.6, 2007.

[216] D.W. Shaw, X. Zhu, M. Misra, R.H. Essenhigh, Determination of global kinetics of coal volatiles combustion, in: 23rd Symposium (International) on Combustion, 1991, pp. 1155–1162.

[217] F.L. Dryer, I. Glassman, High temperature oxidation of CO and CH_4, in: 14th Symposium (International) on Combustion, 1973, pp. 987–1003.

[218] J. Smart, D. Morgan, P. Roberts, The Effect of scale on the performance of swirl stabilised pulverised coal burners, in: 24th Symposium on Combustion, 1992, pp. 1362–1372.

[219] R. Weber, Scaling characteristics of aerodynamics, Heat Transfer, and Pollutant Emissions in Industrial Flames, Proceedings of the Combustion Institute 26 (1996) 3343–3354.

[220] T. Kiga, S. Takano, N. Kimura, K. Omata, M. Okawa, T. Mori, M. Kato, Characteristics of PC combustion in the system of oxygen/recycled flue gas combustion, Energy Conversion and Management 38 (1997) 129–134.

[221] D.W. Pershing, J.O.L. Wendt, Relative contributions of volatile nitrogen and char nitrogen to NO_x emissions from pulverized coal flames, Industrial Engineering Chemistry Process Design 18 (1) (1979) 60–67.

[222] S. Hill, L.D. Smoot, Modeling of nitrogen oxides formation and destruction in combustion systems, Progress in Energy and Combustion Science 26 (2000) 417–458.

[223] H. Stadler, D. Toporov, M. Förster, R. Kneer, On the influence of the char gasification reactions on NO formation in flameless coal combustion, Combustion and Flame 156 (9) (2009) 1755–1763, http://dx.doi.org/10.1016/j.combustflame.2009.06.006.

[224] T.C. Williams, Ch.R. Shaddix, R.W. Schefer, Effect of syngas composition and CO_2-diluted oxygen on performance of a premixed swirl-stabilized combustor, Combustion Science and Technology 180 (2008) 64–88, http://dx.doi.org/10.1080/00102200701487061.

[225] S. Hjärtstam, K. Andersson, F. Johnsson, B. Leckner, Combustion characteristics of lignite-fired oxy-fuel flames, Fuel 88 (11) (2009) 2216–2224.

[226] J. Giménez-López, A. Millera, R. Bilbao, M.U. Alzueta, HCN oxidation in an O_2/CO_2 atmosphere: an experimental and kinetic modeling study, Combustion and Flame 157 (2010) 267–276, http://dx.doi.org/10.1016/j.combustflame.2009.07.016.

[227] J.P. Spinti, D.W. Pershing, The fate of char-N at pulverized coal conditions, Combustion and Flame 135 (2003) 299–313, http://dx.doi.org/10.1016/S0010-2180(03)00168-8.

[228] Y.Q. Hu, N. Kobayashi, M. Hasatani, The reduction of recycled-NO_x in coal combustion with O_2/recycled flue gas under low recycling ratio, Fuel 80 (2001) 1851–1855.

[229] B. Dhungel, P. Mönckert, J. Maier, G. Scheffknecht, Investigation of oxy-coal combustion in semi-technical test facilities, in: Third International Conference on Clean Coal Technologies for our Future, Cagliari, Italy, May 2007, pp. 15–17.

[230] S. Schefer, B. Bonn, Hydrolysis of HCN as an important step in nitrogen oxide formation in fluidised combustion. Part 1. Homogeneous reactions, Fuel 79 (2000) 1239–1249.

[231] M. Ikeda, D. Toporov, D. Christ, H. Stadler, M. Foerster, R. Kneer, Trends in NO_x emission during pulverised fuel oxy-fuel combustion, Energy Fuels 26 (6) (2012) 3141–3149, http://dx.doi.org/10.1021/ef201785m.

A Appendix

A.1 Global char reaction rates

A.1.1 Char-CO$_2$ reaction rates

Half Order Model according to Mann and Kent [195]

Reaction rate:

$$\frac{dm_C}{dt} = -\pi \cdot d_P^2 \cdot k_{kin,CO_2} \cdot \frac{\sqrt{k_{kin,CO_2}^2 + 4 \cdot p_{CO_2} \cdot k_{diff,CO_2}^2} - k_{kin,CO_2}}{2 \cdot k_{diff,CO_2}}$$

$$k_{kin,CO_2} = k_{0,kin,CO_2} \cdot \exp\left(-\frac{E_A}{R \cdot T_P}\right)$$

$$k_{diff,CO_2} = \frac{k_{0,diff,CO_2}}{d_P} \cdot \left(\frac{T_G + T_P}{2}\right)^{0,75}$$

Parameters:

$$k_{0,kin,CO_2} = 7,32 \cdot 10^{-2} \left[\frac{kg}{m^2 \cdot s \cdot Pa^n}\right]$$

$$k_{0,diff,CO_2} = 1 \cdot 10^{10} \left[\frac{kg \cdot m}{m^2 \cdot s \cdot Pa \cdot K}\right]$$

$$E_A = 115 \cdot 10^6 \left[\frac{J}{kmol}\right]$$

First Order Model according to Mayers [196]

Graphite

Reaction rate:

$$\frac{dm_C}{dt} = -\pi \cdot d_P^2 \cdot k \cdot p_{CO_2}^n \quad \text{mit} \quad k = \frac{k_{kin} \cdot k_{diff}}{k_{kin} + k_{diff}}$$

$$k_{kin} = k_{0,kin} \cdot \exp\left(-\frac{E_A}{R \cdot T_P}\right)$$

$$k_{diff,CO_2} = \frac{k_{0,diff,CO_2}}{d_P} \cdot \left(\frac{T_G + T_P}{2}\right)^{0,75}$$

Combustion of Pulverised Coal in a Mixture of Oxygen and Recycled Flue Gas. http://dx.doi.org/10.1016/B978-0-08-099998-2.00019-9

Parameters:

$$k_{0,diff,CO_2} = 1 \cdot 10^{10} \left[\frac{kg \cdot m}{m^2 \cdot s \cdot Pa \cdot K^{0,75}} \right] \qquad n = 1 \quad [-]$$

	T ≤ 1273 K	T > 1273 K
$k_{0,kin} \left[\frac{kg}{m^2 \cdot s \cdot Pa} \right]$	$1,35 \cdot 10^{-4}$	$6,35 \cdot 10^{-3}$
$E_A \left[\frac{J}{kmol} \right]$	$135,53 \cdot 10^6$	$162,13 \cdot 10^6$

Nth-order DL model according to Kajitani et al. [197]
Chinese coal (DL coal)
Reaction rates:

$$\frac{dX_C}{dt} = k \cdot p_{CO_2}^n \cdot (1 - X_C) \cdot \sqrt{1 - \Psi \ln (1 - X_C)}$$

$$\frac{dm_C}{dt} = -(m(t) - m_{ash}) \cdot \frac{dX_C}{dt}$$

$$= -(m(t) - m_{ash}) \cdot k \cdot p_{CO_2}^n \cdot (1 - X_C) \cdot \sqrt{1 - \Psi \ln (1 - X_C)}$$

$$k = k_0 \cdot \exp \left(-\frac{E_A}{R \cdot T_P} \right)$$

Parameters:

$$k_0 = 1,12 \cdot 10^8 \left[\frac{1}{s \cdot MPa^n} \right] \qquad E_A = 240 \cdot 10^6 \left[\frac{J}{kmol} \right]$$

$$n = 0,48 [-] \qquad \Psi = 1 [-]$$

Nth order Roto model according to Ahn et al. [194]
Indonesian sub-bituminous coal
Reaction rates:

$$\frac{dX_C}{dt} = k \cdot p_{CO_2}^n \cdot p_{total}^m \cdot (1 - X_C)^{2/3}$$

$$\frac{dm_C}{dt} = -m_0 \cdot \frac{dX_C}{dt}$$

$$= -m_0 \cdot k \cdot p_{CO_2}^n \cdot p_{total}^m \cdot (1 - X_C)^{2/3}$$

$$k = k_0 \cdot \exp \left(-\frac{E_A}{R \cdot T_P} \right)$$

Parameters:

$$k_0 = 174,1 \left[\frac{1}{s \cdot MPa^{n+m}} \right] \qquad E_A = 71,5 \cdot 10^6 \left[\frac{J}{kmol} \right]$$

$$n = 0,40 [-] \qquad m = -0,65 [-]$$

Nth order Montana Rosebud model according to Dershowitz [200]
Rosebud coal from the US state of Montana
Reaction rates:

$$\frac{dX_C}{dt} = r_m(t) = k \cdot p_{CO_2}^n \quad \text{mit} \quad k = k_0 \cdot \exp\left(-\frac{E_A}{R \cdot T_P}\right)$$

$$\frac{dm_C}{dt} = -m(t) \cdot \frac{dX_C}{dt} = -m(t) \cdot k_0 \cdot \exp\left(-\frac{E_A}{R \cdot T_P}\right) \cdot p_{CO_2}^n$$

Parameter:

$$n = 0,26[-] \qquad E_A = 97,07 \cdot 10^6 \left[\frac{J}{kmol}\right]$$

	$k_0 \left[\frac{1}{s \cdot Pa^n}\right]$	
T [K]	$p_{CO_2} = 0,26\ atm$	$p_{CO_2} = 0,58\ atm$
1473	$1,05 \cdot 10^2$	$1,5 \cdot 10^2$
1773	$1,05 \cdot 10^2$	$8,6 \cdot 10^1$
2113	$7,4 \cdot 10^1$	$7 \cdot 10^1$

L-H DL model according to Kajitani et al. [197]
Chinese coal (DL coal)
Reaction rates:

$$\frac{dX_C}{dt} = k \cdot S \qquad S = S_0 \cdot (1 - X_C) \cdot \sqrt{1 - \Psi \ln(1 - X_C)}$$

$$k = \frac{k_1 \cdot p_{CO_2} + k_4 \cdot p_{CO_2}^2}{1 + k_2 \cdot p_{CO_2} + k_3 \cdot p_{CO}} \qquad k_i = k_{0,i} \cdot \exp\left(-\frac{E_{A,i}}{R \cdot T_P}\right)$$

$$\frac{dm_C}{dt} = -(m(t) - m_{Asche}) \cdot \frac{dX_C}{dt}$$

$$= -(m(t) - m_{Asche}) \cdot S_0 \cdot (1 - X_C) \cdot \sqrt{1 - \Psi \ln(1 - X_C)}$$

$$\cdot \frac{k_1 \cdot p_{CO_2} + k_4 \cdot p_{CO_2}^2}{1 + k_2 \cdot p_{CO_2} + k_3 \cdot p_{CO}}$$

Parameters:

$$k_{0,1} = 1,278 \left[\frac{g}{m^2 \cdot s \cdot Pa}\right] \qquad E_{A,1} = 222 \cdot 10^6 \left[\frac{J}{kmol}\right]$$

$$k_{0,2} = 3,168 \cdot 10^{-6} \left[\frac{1}{Pa}\right] \qquad E_{A,2} = -23 \cdot 10^6 \left[\frac{J}{kmol}\right]$$

$$k_{0,3} = 1,273 \cdot 10^{-6} \left[\frac{1}{Pa}\right] \qquad E_{A,3} = -48,1 \cdot 10^6 \left[\frac{J}{kmol}\right]$$

$$k_{0,4} = 2,146 \cdot 10^{-16} \left[\frac{g}{m^2 \cdot s \cdot Pa^2} \right] \qquad E_{A,4} = 68,5 \cdot 10^6 \left[\frac{J}{kmol} \right]$$

$$S_0 = 190 \left[\frac{m^2}{g} \right] \qquad \qquad \Psi = 1 [-]$$

A.1.2   Char-steam reaction rates

Nth Montana Rosebud according to Dershowitz [200]
   Rosebud coal from the US state of Montana.
   Reaction rates:

$$\frac{dX_C}{dt} = r_m(t) = k_0 \cdot p_{H_2O}^n \qquad \text{mit} \quad k = k_0 \cdot \exp\left(-\frac{E_A}{R \cdot T_P}\right)$$

$$\frac{dm_C}{dt} = -m(t) \cdot \frac{dX_C}{dt} = -m(t) \cdot k_0 \cdot \exp\left(-\frac{E_A}{R \cdot T_P}\right) \cdot p_{H_2O}^n$$

Parameters:

$$n = 1,19[-] \qquad E_A = 109,62 \cdot 10^6 \left[\frac{J}{kmol} \right]$$

	$k_0 \left[\frac{1}{s \cdot Pa^n} \right]$	
T [K]	$p_{H_2O} = 0,335 \, atm$	$p_{H_2O} = 0,65 \, atm$
1473	$7 \cdot 10^{-3}$	$12 \cdot 10^{-3}$
1773	$16 \cdot 10^{-3}$	$12 \cdot 10^{-3}$
2113	$10 \cdot 10^{-3}$	$5,5 \cdot 10^{-3}$

B Appendix

B.1 Plug flow reactor program

See Figure B.1.

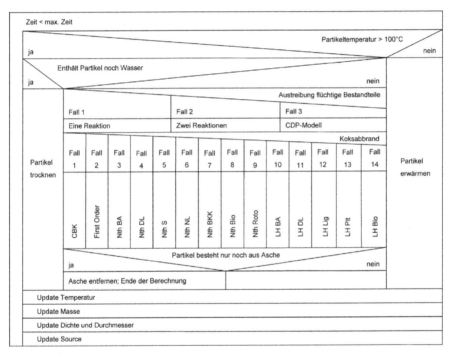

Figure B.1 Plug flow reactor algorithm.

Combustion of Pulverised Coal in a Mixture of Oxygen and Recycled Flue Gas. http://dx.doi.org/10.1016/B978-0-08-099998-2.00020-5

Printed and bound by CPI Group (UK) Ltd, Croydon, CR0 4YY

03/10/2024

01040413-0014